The Climate Trial

Saúl Luciano Lliuya in the Andes. (Photo: Alexander Luna)

The Climate Trial

LAW AND JUSTICE ON A MELTING PLANET

Noah Walker-Crawford

DUKE UNIVERSITY PRESS
Durham and London 2026

Cover design by Dave Rainey
Typeset in Minion Pro and Real Head Pro by Westchester Publishing Services

Library of Congress Cataloging-in-Publication Data
Names: Walker-Crawford, Noah, [date] author.
Title: The climate trial : law and justice on a melting planet / Noah Walker-Crawford.
Other titles: Law and justice on a melting planet
Description: Durham : Duke University Press, 2026. | Includes bibliographical references and index.
Identifiers: LCCN 2025028355 (print)
LCCN 2025028356 (ebook)
ISBN 9781478033172 (paperback)
ISBN 9781478029724 (hardcover)
ISBN 9781478061915 (ebook)
ISBN 9781478094487 (ebook other)
Subjects: LCSH: RWE (Firm : 1990–)—Trials, litigation, etc. | Climate justice—Germany. | Environmental justice—Germany. | Climatic changes—Law and legislation—Germany. | Liability for climatic change damages—Germany. | Environmental responsibility—Germany. | Climatic changes—Social aspects—Peru. | Glaciers—Peru. | Climate change mitigation—Peru. | Indigenous peoples—Peru—Environmental conditions. | Peru—Environmental conditions.
Classification: LCC GE240.G3 W355 2026 (print) | LCC GE240.G3 (ebook) | DDC 363.7/05610943—dc23/eng/20260114
LC record available at https://lccn.loc.gov/2025028355
LC ebook record available at https://lccn.loc.gov/2025028356

Cover art: Eduardo Díaz addresses the mountains. Photo by Alexander Luna, 2026. Courtesy of the artist.

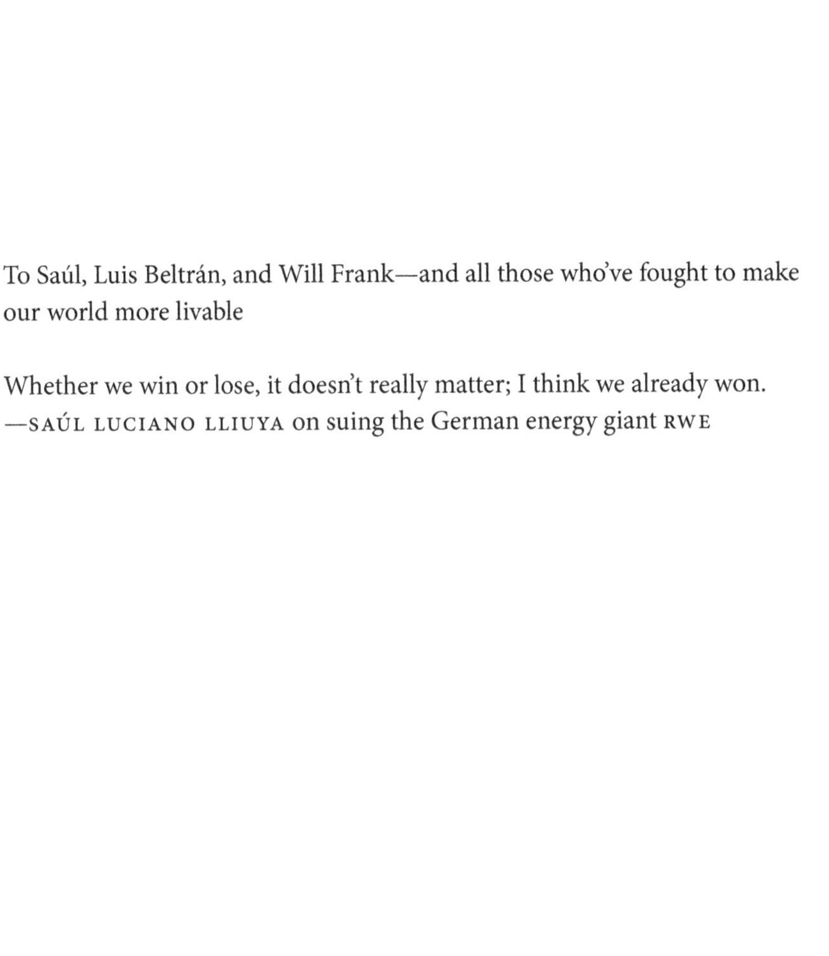

To Saúl, Luis Beltrán, and Will Frank—and all those who've fought to make our world more livable

Whether we win or lose, it doesn't really matter; I think we already won.
—SAÚL LUCIANO LLIUYA on suing the German energy giant RWE

Contents

MAP 1. Map of Peru. (Source: D-Maps, https://d-maps.com/carte.php?num_car=15163)

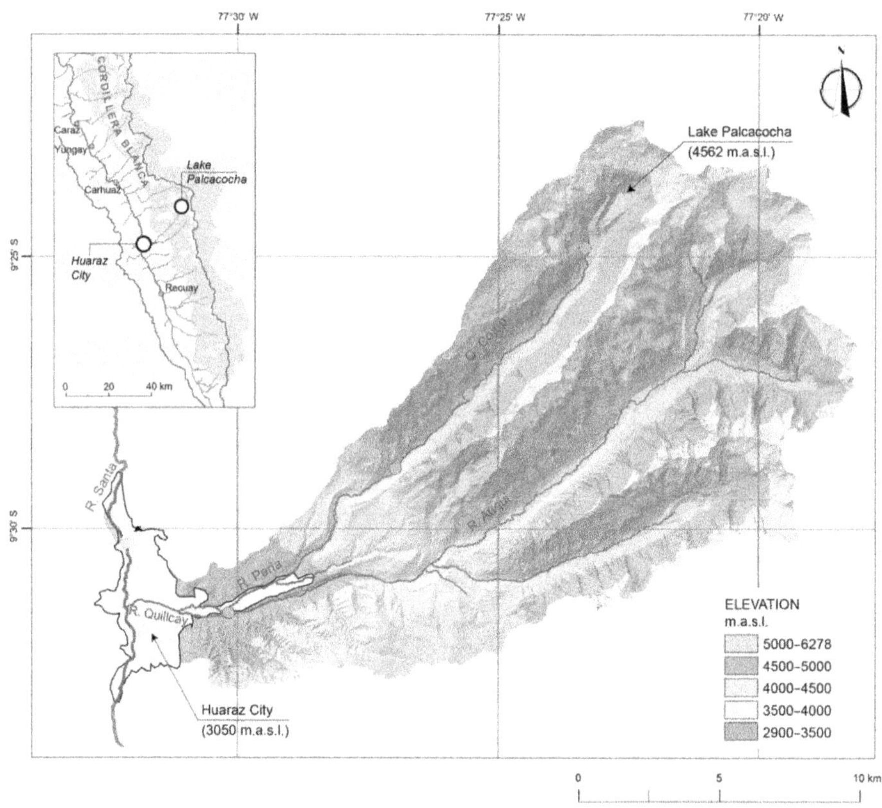

MAP 2. Huaraz and Lake Palcacocha. (Source: Wiki Commons, https://commons.wikimedia.org/wiki/File:SomosValenzuelaHESS2016-fig1-mode-en.png.

Abbreviations and Glossary

AR5
Fifth Assessment Report

GLOF
glacial lake outburst flood

INAIGEM
National Research Institute for Glaciers and Mountain Ecosystems

IPCC
Intergovernmental Panel on Climate Change (United Nations body for climate science)

NGO
nongovernmental organization

RWE
Major German energy corporation founded in 1898; known as Rheinisch-Westfälisches Elektrizitätswerk (Rhenish-Westphalian Electricity Works) until 1990

UNFCCC
United Nations Framework Convention on Climate Change

Essen State Court (Landgericht Essen)
Court of first instance for *Luciano Lliuya v.* RWE

Germanwatch
German environmental NGO backing the plaintiff in *Luciano Lliuya v.* RWE

Glacier Authority
Glacier and Lake Evaluation Area (Área de Evaluación de Glaciares y Lagunas); governmental monitoring agency in Huaraz

Upper State Court, Hamm (Oberlandesgericht Hamm)
Appeals court hearing *Luciano Lliuya v.* RWE

Key Characters

Note: The names of publicly known people have not been changed. This includes prominent characters with a role in the lawsuit between Luciano Lliuya and RWE. Other names have been changed to preserve anonymity.

Martín Amaru (name changed): Worker at the glacial lake safety project at Palcacocha

Christoph Bals: Policy director at Germanwatch

Eduardo Díaz: Foreman of the glacial lake safety project at Palcacocha

Saúl Luciano Lliuya: Plaintiff in the case of *Luciano Lliuya v. RWE*

Julio Luciano Shuan: Saúl Luciano Lliuya's father

Rolf Meyer: Head judge in the trial of *Luciano Lliuya v. RWE* at the Upper State Court in Hamm

Klaus Milke: Cofounder and former chairman of the board at Germanwatch

Pedro Vasquez (name changed): Government-employed engineer involved in glacial lake monitoring in the Cordillera Blanca

Roda Verheyen: Environmental lawyer representing the plaintiff in *Luciano Lliuya v. RWE*

Fernando Vilca (name changed): Engineer overseeing the glacial lake safety project at Palcacocha

Introduction

Climate Justice in Court

"Today, the mountains have won."[1]

Saúl Luciano Lliuya's words reverberated around the building of Upper State Court in Hamm, captivating a crowd of people and TV cameras. He had come a long way from his home in the Peruvian Andes to lead a historic legal case in Germany, and the judges had just found it admissible. It seemed outlandish: A small-scale farmer from a Quechua-speaking community took the German energy giant RWE to court over its contribution to climate change impacts in the Andes. RWE had no operations in Peru but had produced a substantial amount of emissions through operating coal-fired power plants in Europe for over a century. Luciano Lliuya's community has faced dramatic changes in the mountain environment. The plaintiff worked as a mountain-climbing guide, a job that brought him into contact with glaciers that are rapidly disappearing. In the long run, many locals were concerned about water scarcity caused by melting glaciers. In the short term, there could be too much water: Glacial retreat has caused mountain lakes to grow to unstable levels, raising the risk of flooding for downstream communities. Luciano Lliuya owns a house that lies below the lake known as Palcacocha, which scientists have identified as liable to overflow its banks. In the German courtroom, Luciano Lliuya sought to hold RWE liable for its contribution to flood risk in Peru and make the company contribute around US$20,000 to stabilize the lake.

The argument was simple: climate change makes us all neighbors on our shared planet. As neighbors, we have the responsibility to act kindly toward one another. Major fossil fuel companies like RWE made a substantial contribution to climate change, dramatically transforming Luciano Lliuya's life in Peru. RWE should be a good neighbor and help him deal with the consequences of the company's actions. Filed in 2015 with support from the German nongovernmental organization (NGO) Germanwatch, the claim sought to set a massive precedent to hold major polluters responsible for climate change. While a lower court dismissed the case, it was now on appeal in the Upper State Court in Hamm. In a courtroom packed full of journalists and climate activists in November 2017, a panel of three German judges came to a surprising conclusion: the case was solid. They agreed that neighborhood law could be applied to climate change, meaning that the case was admissible. In principle, they saw no reason why a major polluter should not be held responsible for its contribution to global warming. The case would proceed to the evidentiary stage, and if the court found enough proof linking RWE's emissions to glacial retreat and flood risk in Peru, Luciano Lliuya would likely win.

Leaving the monumental hearing with his lawyers, Luciano Lliuya expressed surprise and was overwhelmed by emotion. It was the furthest any case of this kind had ever gotten. Standing by his side outside the courtroom, I glanced around the crowd and saw many members of the audience close to tears. In Luciano Lliuya's mind, he pictured the mountains in Peru. The mountains had pushed him to take a stand in the German courtroom. For this cause, he had traveled the world and faced his fear of public attention. Luciano Lliuya is not a tall man. The cameras and journalists stared down at him (figure I.1). He thought about his grandparents' stories of the Andean glacial lakes. The mountains were now suffering. He was at court with RWE to fight for them: "The lakes are the tears of the mountains. Today, justice heard the mountains crying."[2]

The case began with an unlikely encounter: In 2014, following the COP20 United Nations Climate Summit in Lima, a team from the NGO Germanwatch traveled to the Cordillera Blanca region in the Peruvian Andes. The area had faced accelerated glacial retreat in recent decades, making it an international climate change hotspot. Luciano Lliuya was concerned about these changes, which threaten his community's agricultural livelihood. He became familiar with the discourses of global climate change through the media and discussions with foreign mountain-climbing tourists. He came to recognize that blame for the changes lies not with Peruvians like himself but with wealthy countries and large companies that have caused most global greenhouse gas emissions. Luciano Lliuya met the group of German climate activists through a mutual friend. They were part of international efforts to develop legal tools for addressing climate change. Political negotiations at UN Climate Summits were faltering, and they sought to place pressure on industry and governments. At that point, no lawsuit had successfully held a company or country liable for its contribution to climate change. After extended

FIGURE I.1. Luciano Lliuya speaking to the press following his court hearing in Hamm, Germany, November 2017. To left stands his lawyer, Roda Verheyen; to the right are author, Germanwatch cofounder Klaus Milke and Germanwatch policy director Christoph Bals. (Photo: Alexander Luna)

discussions, the German activists offered Luciano Lliuya the possibility of making a claim against a major European emitter. Now he was fighting his case in court.

During the 2017 hearing, head judge Rolf Meyer explained the court's holding that the case was legally admissible. RWE's lawyers quickly rebuffed the judge's suggestion of an out-of-court settlement, arguing that this case would set a precedent. For the climate activists at Germanwatch who had collected donations to organize the lawsuit, this was a legal test case: Could nuisance law be used to hold major corporations liable for their contribution to climate change? Similar cases in other jurisdictions had failed,[3] but climate science had evolved rapidly, improving the evidentiary basis for linking emitters and impacts. This case addressed a significant injustice of climate change: While most emissions have been produced in wealthy countries of the Global North, many of the worst impacts are felt in the Global South where people and governments have fewer resources needed to respond. Bringing this moral dimension to the forefront, Judge Meyer remarked during the hearing: "But in the places in the world where money is scarce, can we leave these people on their own even when we are causing the problem over here? *Is that just*?"[4]

The judge's questioning—*Is that just?*—was surprisingly broad given the lawsuit's limited subject matter. Formally, it concerned a relationship between Saúl Luciano Lliuya, one of around fifty thousand people living in a flood-prone area in the Peruvian Andes, and RWE, a corporate legal person based in Germany. At stake was the question whether RWE was partially responsible for glacial lake outburst flood risk affecting Luciano Lliuya's property. Drawing on a study that quantified individual companies' contributions to global warming (Heede 2014a), the lawsuit alleged that RWE caused 0.47 percent[5] of industrial greenhouse gas emissions and should cover 0.47 percent of the cost for a government project to secure Palcacocha, equating to around US$20,000. That sum was symbolic as the legal costs were much higher. The judge's moral deliberations made explicit what all those involved in the legal process already knew: This lawsuit was about much more than a private nuisance claim between Luciano Lliuya and RWE. It raised fundamental questions about who should take responsibility for climate change.

Climate change is not just transforming the Earth's physical environment; it changes how we all relate to one another. This book tells the inside story of a climate justice lawsuit that shook the world, and reflects on what it means for how we should live together on our warming planet. We will travel between the Peruvian Andes, German courts, and UN Climate Summits; we will follow legal arguments, trace the production of scientific evidence, and confront political controversy. On the fringes we encounter powerful Andean mountains that have no standing in the courtroom yet shape legal and political disputes in peculiar ways.

My involvement in the lawsuit began when I worked for Germanwatch and helped coordinate the claim in its initial stages. I went on to study the case as an ethnographic researcher and continued to support Luciano Lliuya and his legal team. I share with you a perspective from the plaintiff's side on how the case emerged, and I use the lawsuit as a starting point to explore the moral stakes of climate change. The claim connected Luciano Lliuya and RWE by defining them as neighbors. Climate activists drew on the idea of neighborliness to argue for the responsibility of major emitters to address the devastating impacts of global warming. Building on this approach, I use neighborliness as an analytical framework to study the social and moral dynamics of climate change. Following a dispute between two unlikely neighbors, this book uncovers how climate change reshapes moral relations across the planet.

Climate Change Is a Neighborly Dispute

The material processes of climate change are well understood. Climate models demonstrate how greenhouse gases become trapped in the Earth's atmosphere, leading to a warming effect on the planet's surface. This contributes to a wide variety of phenomena including sea level rise, glacial retreat, and deadly heat waves. Extensive scientific research demonstrates how climate change impacts people's lives, exacerbating existing vulnerabilities and generating unprecedented dangers. Atmospheric models illustrate the physical dynamics, but they do not say how we should deal with the consequences. To make arguments about who should take responsibility for climate change, people link scientific representations of global warming to moral conceptions regarding how people, institutions, and environments should engage with one another. One such approach is the legal argument that climate change makes us all neighbors.

Neighbors are actors with mutual moral obligations, often arising out of physical or conceptual proximity. The term *neighborliness* and the adjective *neighborly* refer to ideas about how neighbors should rightfully treat one another. Neighborliness is a familiar moral framework that resonates with people around the world. I distinguish between normative and analytical conceptions of neighborliness: In a normative sense, appeals to neighborly relations posit that people should act in a certain way toward one another. Those are the claims I study in this book. I begin with neighborliness in examining the moral dynamics of climate change from an anthropological perspective. The analytic of neighborliness is a tool for studying moral relations across local and global scales. Here I study how moral relationships are constructed between humans, corporations, and other actors. This approach highlights a fundamental ambiguity at the heart of legal climate justice claims: They appeal simultaneously to individual and collective moral responsibility.

The lawsuit between Luciano Lliuya and RWE drew on legal norms that people usually invoke to seek relief from neighbors for damage or potential harm to their property. In their arguments, Luciano Lliuya's lawyers expanded the legal conception of neighborliness to encompass relations across the planet: As climate change has connected RWE and Luciano Lliuya, it has made them neighbors. Law codifies who counts as a neighbor and what constitutes good neighborly behavior. Strictly speaking, Luciano Lliuya's claim concerned the relationship between two legal persons. It individualized climate change by framing it as a dispute between one human and one company. Yet, for Luciano Lliuya and his supporters, the lawsuit was also an attempt to set a precedent that would govern relations between all polluting corporations and all people affected by climate change.

Beyond the legal framework, the lawsuit provided a platform for its proponents to make broader normative arguments about who should take responsibility for climate change. This issue concerns social relations among countless people who face the devastating impacts of climate change, numerous corporations, and governments that continue to promote a fossil fuel–based economic model, as well as the Earth itself. The normative appeal to neighborliness addresses the question of how we should live together on our planet. This approach links an everyday understanding of neighborliness, as relations between people who live close to one another, to the legal definition—legal persons shouldn't interfere with each other's property—and expands the concept to a global level. Luciano Lliuya and his NGO backers broaden the moral basis of neighborliness beyond property rights, arguing that neighbors should act in a positive way toward each other and should not cause each other harm. They make the universalizing moral claim that climate change makes us all neighbors—all humans and corporations. This premise allows them to argue that fossil fuel companies, which have caused harm to others through their contribution to climate change, should take responsibility and provide redress to those harmed because they have acted as bad neighbors.

The normative conception of neighborliness deployed in and around the lawsuit against RWE involves a fundamental ambiguity: It individualizes climate change by framing it in terms of relations between specific actors, and it universalizes those relations by claiming that we should all be good neighbors. It involves claims not only about how individuals should interact with each other but also about how social relations should be governed more broadly—in the global neighborhood. The appeal to neighborliness simultaneously individualizes and collectivizes the issue of climate change. This ambiguity lends the concept strength: The idea of neighborly relations is easily understandable to anyone who has lived in a community, and it allows people to draw the imaginative link from the individual to the collective scale.

The aim of this book is not to make a moral argument. Rather, I examine how social relations are redefined in response to climate change, and how climate change reshapes moral relations across local and global scales. I use the concept of neighborliness to address these questions from an anthropological perspective. I do not argue *that* climate change makes us all neighbors;

I explore *how* a neighborliness perspective allows for a simultaneous appeal to individual and collective responsibility. I briefly outline anthropological discussions of climate change, highlighting how the neighborliness analytic offers a new perspective.

Anthropology is the study of human behavior and social life. Its principal research method is ethnography, which involves documenting people's social interactions and following how they engage with life's challenges. Anthropology offers a unique view for studying how climate change transforms social life. This connection covers several broad aspects (O'Reilly et al. 2020). First, anthropologists have traced the production and circulation of climate change knowledge, focusing in particular on science. This focus provides a critical view of how climate change debates are shaped by politics, power dynamics, and cultural values (Barnes et al. 2013). Second, ethnographic research shows how people engage with climate change impacts. Collaborating with natural scientists, anthropologists have contributed sociocultural approaches to climate change research (Crate 2011). Finally, anthropologists have examined efforts to mitigate the climate crisis, with a significant focus on the transition from fossil fuels to renewable energy (High and Smith 2019). Many of these studies involve a territorially situated examination of climate change knowledges, impacts, and mitigation strategies. Grounded in a tradition of long-term ethnographic fieldwork, anthropology is a useful tool for studying how people engage with climate change in different places around the world. Some researchers have begun to theorize from an anthropological standpoint how climate change connects people, the planet, and its atmosphere. Knox (2020) suggests that we approach climate change via its material processes in order to study the energy relations it invokes. In this way, we can trace the material and social relations through which climate change emerges and becomes contested. Following the suggestion of O'Reilly and colleagues (2020, 23) that anthropology can help us "reimagine the future of human-atmosphere relations," I study climate change ethnographically by tracing how its material relations give rise to new moral relationships.

What makes relations moral? *Morality* and *ethics* refer to normative concerns about how people should act.[6] While philosophical discussions often revolve around defining ethical principles, an ethnographic perspective can trace how normative concerns emerge through people's lived experience and how they navigate moral tensions (Fassin 2012). Moving between scales of analysis, anthropological study relates ethnographic experiences of moral issues to broader analytical and social concerns in daily life (Fassin 2011).

Such research can show how climate change arises as a moral concern in public debates and activist practice (von Storch et al. 2021). Ethical norms about how social actors should engage with one another are embedded in legal institutions. Law consists of codified rules and moral standards, shapes categories of identity, and can serve as a framework for political action (Goodale 2017, 4). While law has often been used as a tool of oppression, subaltern groups can mobilize legal tools to challenge the status quo (Eckert et al. 2012). Ethnographic study of judicial process traces how legal concepts are negotiated in practice. Exploring the analytical implications of legal debates, such research can inform broader understandings in social theory about the issues at stake (Bens and Vetters 2018).

Countless studies examine engagements between neighbors at a local level (e.g., Henig 2012; Zabiliūtė 2020) or between adjacent ethnic and national populations (e.g., Åtland 2010; Gribetz 2014). Other academics use the term *neighbor* in a metaphorical sense to discuss moral relations between people around the world. Some draw on a Christian ethics of neighborliness to promote more charitable engagement between people in a globalized world (Walker 2008). Others invoke the idea of "global neighbors" to promote ethical consumerism addressing global inequality (Haugestad 2004). The "global neighbor" idiom has a clear analytical value: it posits the centrality of moral relations. However, I define *neighborly relations* in more concrete ethnographic terms as moral relations between social actors who are able to affect each other. Appeals to neighborliness simultaneously invoke ideas about individual and collective relations. Saying that someone should be a good neighbor involves a moral claim about how that person should act toward you and about how people should treat each other in the neighborhood.

Neighborliness is a contested terrain. Its bounds are negotiated in legal and moral disputes about what responsibilities people and other actors should have toward one another. I distinguish neighborliness from kinship, which involves a much tighter set of relations. Picking up on long-standing debates in anthropology, Sahlins (2013) has defined kinship as the "mutuality of being": It involves a social interdependence at an existential level. Based on a review of numerous ethnographic studies from around the world, Sahlins argues that kinspeople make one another who they are because they participate in one another's lives. Neighborliness is broader in scope: It involves those who are not close enough to be kin yet are not strangers. In philosophical discussions about the ethics of coexistence, the "neighbor" has been theorized as a figure toward whom one has potential moral obligations (Thiranagama 2019). Rather than mutuality of being, neighbors are held together through mutual responsibility. This bond can be regulated through law and local custom, and the bounds of responsibility are often disputed. Neighborly relations can involve numerous types of social actors, including humans, corporations, and sentient ecosystems (de la Cadena 2015). Neighborly relations are not necessarily harmonious; they can be antagonistic and conflictual. A person who asserts that someone else is a neighbor, calls on the other to accept the ensuing legal and moral responsibility.

Our increasingly connected world gives rise to new kinds of moral relations. Issues like climate change, migration, and global trade make it clear that the actions of one individual or entity can have far-reaching consequences for others across the globe. This fact raises profound questions about how moral responsibility is distributed among individuals, corporations, states, and other actors. Some promote a universalist notion of cosmopolitanism and global citizenship, positing that all humans are members of a single global community, bound by universal moral obligations derived from a shared humanity. In this framework, all individuals are entitled to the same rights and bear the same responsibilities to one another, regardless of geographic or national boundaries (Appiah 2005; Beck 2006).

Others have critiqued this universalist vision of moral relations, particularly as it can obscure power relations. By treating all individuals or nations as equal members of a global moral community, these frameworks may overlook historical inequalities—such as those between the Global North and Global South—that continue to shape contemporary relationships (Fraser 2009; D. Harvey 2009). In addition, such frameworks have been critiqued for being too focused on humans, neglecting the broader ecological relationships and responsibilities that arise in the context of planetary crises like climate change (Latour 2018). Ideas of cosmopolitanism often disregard moral relationships with other species, ecosystems, and the planet itself.

Building on these critiques, Bruno Latour (2021) offers an alternative approach rooted in relational ethics. He argues that the challenge of global moral responsibility cannot be adequately addressed through universalist principles alone. Instead, he suggests that we should understand ourselves as Earthbound, emphasizing our deep entanglement with one another and with the planet's ecological systems. From this perspective, moral responsibility stems not from abstract principles of shared humanity but from the specific relationships and interdependencies that link humans, nonhumans, and ecosystems. For Latour, global problems like climate change demand a moral framework that recognizes these contextual and relational ties, rather than imposing a one-size-fits-all standard of moral obligation.

This book traces how moral relations are constructed across geographic and conceptual scales: from the local to the global; from the individual to the universal. The concept of neighborliness recognizes that moral claims about the nature of global relations can be both specific and contextual, as in Latour's view, and universalistic, as in the cosmopolitan vision. Neighborly obligations can arise from specific, situated interactions while also pointing to broader, universal principles of justice and responsibility. The neighborliness analytic shows who is included in claims about neighborliness, and who is left out. A focus on this ambiguity between claims of particularity and universality can account for the ways power relations shape moral engagements around climate change, and considers the role of other potential actors such as corporations and ecosystems. Applied to climate litigation, the neighborliness analytic uncovers the complexities, nuances, and contradictions of global claims about who should take responsibility for climate change. It highlights a tension between individual experiences of climate change and the collective human and planetary plight that is captured in scientific research and modeling. It points to the ambiguity at the heart of claims about who is responsible: They invoke relations between individual actors—RWE and Luciano Lliuya—and universal relations between all polluters and all those affected.

Neighborly relations are rooted in shared locality, and climate change ties localities together across the planet. This book explores how the concept of neighborliness helps us understand the social and moral stakes of climate change as reflected in claims about individual and collective responsibility. Approaching climate change in terms of neighborly relations makes it possible to study the issue ethnographically. Anthropological research can follow discussions about causal linkages and moral responsibility in scientific, legal, and political contexts. Addressing climate change as an issue of neighborly relations makes its moral dimensions methodologically and theoretically amenable. The lawsuit between Luciano Lliuya and RWE provides an empirical opportunity to study how people invoke moral bonds to make sense of climate change. Global warming is a fundamentally moral issue: It expands the scope of social relations and asks how we should live together on our planet.

The Social Life of Climate Law

Global leaders in government and business have been slow to respond to the climate crisis. Citizens around the world are increasingly taking the issue to court, as I explain in chapter 2. Luciano Lliuya's claim against RWE was one of almost three thousand cases filed worldwide, with around two-thirds brought to court since the Paris Agreement was adopted in 2015 (Setzer and Higham 2025). Since this development, academic literature on climate litigation has flourished. A 2020 review identified almost two hundred articles (Peel and Osofsky 2020); that number likely doubled by 2026. I provide a brief overview of these discussions and highlight my contribution.

Most publications focus on legal aspects of climate litigation. A number of authors have published broad overviews on the development of climate litigation (Mitkidis and Valkanou 2020; Setzer and Higham 2025). Edited volumes have been published with comparative law perspectives (Alogna et al. 2021), on human rights and climate litigation (Rodríguez-Garavito 2022), and climate litigation in the Asia Pacific (Lin and Kysar 2020) and Africa (Bouwer et al. 2024). Legal scholars have written about various themes in and approaches to climate litigation, including human rights (Peel and Osofsky 2017; Beauregard et al. 2021; Iyengar 2023), claims against government about their climate policy ambitions (Minnerop 2022; Hellner and Epstein 2023), claims against major corporate emitters (Ganguly et al. 2018; Bouwer 2020; Verheyen and Franke 2023), and youth claims (Donger 2022; Parker et al. 2022; Wewerinke-Singh and Nay 2023). Some articles focus explicitly on legal theory development, addressing challenges such as suing corporations over climate harms (Kumar and Frank 2018) and establishing legal standing in court for plaintiffs (Kelleher 2022). Numerous articles analyze cases and legal challenges in specific jurisdictions, with a particular focus on notable cases such as *Neubauer v. Germany* (Aust 2022; Ekardt and Heyl 2022; Winter 2022) and *Urgenda Foundation v. State of the Netherlands* (Leijten 2019; Wewerinke-Singh and McCoach 2021) in which citizens successfully forced their governments to take more decisive policy action on climate change. Other work examines the challenges of climate litigation in geographic areas such as the Global South (Setzer and Benjamin 2020; Auz 2022b) and Latin America (Auz 2022a; Cavedon-Capdeville et al. 2024).

This book contributes to an emerging academic dialogue on the social dynamics of climate litigation. It speaks to the law literature by highlighting aspects of climate litigation that are not captured by doctrinal and empirical legal research. Literature on "legal mobilization" shows that litigation is part of broader activist efforts to achieve social and political change (Vanhala 2022a). Such research highlights the sociopolitical and institutional dynamics of climate litigation and points to its impacts beyond the legal sphere (Vanhala 2022b). A number of academics have examined narrative aspects of climate litigation, including how narratives of climate justice are transplanted across borders (Paiement 2020), how litigation serves as a platform for storytelling (Rogers 2020), and how climate litigation influences discussions in the media and politics (Wonneberger 2023; Wonneberger and Vliegenthart 2021). A growing body of work examines the interplay between litigation and climate politics, particularly in discussions at the United Nations about Loss and Damage. Researchers argue that the threat of litigation boosts Global South countries' political demands for increased support and compensation from major emitters (Toussaint 2021; Wewerinke-Singh 2023).

My work (Walker-Crawford 2021, 2023, 2024) contributes an in-depth perspective on the practical dynamics of climate litigation. This book is the first ethnographic monograph on the topic. With some notable exceptions (Geiling 2019; Supran and Oreskes 2021), the overwhelming majority of literature on climate litigation focuses on the perspectives of plaintiffs demanding greater climate action from governments and corporations. To gain a more holistic understanding of climate litigation, we must also examine what defendants have to say. Part 2 of this book provides an in-depth discussion of both Luciano Lliuya's and RWE's arguments about evidence and causation. Overall, I contribute to academic discussions about why climate litigation matters. This book provides an inside view of how an emblematic climate justice claim has played out between melting glaciers, courtrooms, and the global stage of climate politics.

Activism and Engaged Anthropology

I tell this story from the viewpoint of a professional activist–turned–engaged anthropologist. My perspective is necessarily partial: I narrate Luciano Lliuya's lawsuit from within the plaintiff's legal team, drawing on my experience of accompanying him in the Peruvian mountains, German courts, and UN Climate Summits. Throughout this journey, I wear two distinct hats: As an activist, I supported the lawsuit to further the goal of climate justice. As an anthropologist, I reflect on how climate litigation reshapes the moral stakes of climate change. At times, these roles pulled me in different directions. Yet this tension is productive: My involvement in the case gives an insightful ethnographic snapshot. My analysis does not provide all the answers about how the world should deal with climate change, but it informs activist efforts by asking why climate litigation matters. As an engaged anthropologist, I use the tools of academic inquiry to argue for justice in a warming world.

I first joined Germanwatch in 2014 when I was an undergraduate anthropology student. Germanwatch is an environmental NGO with a major focus on climate politics. The organization has been involved in the UN climate negotiations since they began in the 1990s and has lobbied the world's governments to take more effective action on climate change. After attending the 2014 UN Climate Summit in Lima, I traveled to the Andean city of Huaraz, Peru, with a small team from Germanwatch to see the effects of glacial retreat up close. Since I had lived in Peru before going to university and none of the others spoke Spanish, I acted as the group's guide. We met Luciano Lliuya through a friend of mine, an agricultural engineer who worked with farmers in the region. Luciano Lliuya shared his worries about how climate change was affecting his community and expressed frustration that faraway countries and industries are to blame. We told him about the possibility of bringing a legal claim against a major German emitter and later put him in touch with a lawyer. As these discussions began, I acted as Luciano Lliuya's interpreter and confidante.

After Luciano Lliuya decided to move forward with the suit, I worked with Germanwatch and the lawyers to gather evidence and prepare legal arguments linking RWE's activities to glacial retreat and flood risk in Peru. I acted as the contact person between the German team and the plaintiff. In countless conversations, Luciano Lliuya and I talked through legal scenarios and how the case might influence political discussions about climate change.

Working on the lawsuit raised broader questions for me: Who should take responsibility for climate change? How well does climate litigation address this question? What perspective does an Andean standpoint provide? I left my role at Germanwatch to pursue these issues through a PhD in social anthropology at the University of Manchester. I conducted ethnographic fieldwork in Peru for twenty months, studying how rural Andeans engage with climate change. I accompanied Luciano Lliuya to court hearings and UN Climate Summits. All the while, I continued to participate in strategy discussions with Luciano Lliuya's legal team. After finishing my PhD, I began conducting comparative research on the use of climate science as legal evidence and took on a greater role in the lawsuit as a consultant to Germanwatch, providing strategic advice and overseeing the lawsuit's scientific argumentation.

Throughout this journey, I have contributed to the lawsuit in four key areas: as a coordinator, translator, scientific adviser, and public spokesperson. First, I coordinated contact between the legal team in Germany and our Peruvian interlocutors: Luciano Lliuya, local lawyers, campaigners, and community organizers.

Second, I acted as Luciano Lliuya's interpreter in discussions with the legal team. This required more than explaining German judicial concepts in Spanish. It involved a conceptual translation between legal and political activist ideas and Luciano Lliuya's Andean perspective. While the primary aim for Germanwatch was to intervene in global climate politics, Luciano Lliuya's main concern was limiting global warming to stop his mountains from suffering. The two found common ground in the lawsuit with its appeal to neighborly relations. My role was to help Luciano Lliuya and the legal team understand each other's concerns and work together effectively.

My third contribution to the case was as a scientific adviser. This work involved a different type of translation: between scientific and legal frameworks. Science and law use different approaches in deciding whether facts are true and relevant. In this capacity, I conducted research, sought advice from climate change experts, and summarized scientific evidence in a way that made sense to lawyers and judges. I contributed to the plaintiff's legal briefs submitted to the court and participated in court hearings as an expert witness. Part 2 of this book provides an in-depth view of this process, showing how evidentiary discussions played out in the trial.

My final role in the case was as public spokesperson. When the case began, I usually acted as Luciano Lliuya's interpreter during media interviews. When Luciano Lliuya spoke to the crowd outside the courtroom after his hearing in 2017, I translated his words into German. As the case gained traction, it received significant attention from the global press. I increasingly gave interviews to the media myself, speaking both as an academic studying climate litigation and as an adviser to Luciano Lliuya's legal team.

My involvement in the lawsuit necessarily shapes this book's analysis. I offer a perspective on an emblematic climate lawsuit from within the plaintiff's legal team. I do not provide RWE's viewpoint in the same depth. I gain the company's perspective from its legal submissions and its lawyers' statements in court. This book interrogates legal argumentation on both sides, showing how it relates to broader concerns about who should take responsibility for climate change.

Balancing scholarly research and activism is a challenge. The two endeavors involve different types of storytelling: Where activists devise simple stories to motivate social action, academics focus on complexity. Anthropologists are notorious for arguing that everything is more complicated than people think, often leaving their readers with more questions than answers. According to Hastrup and colleagues (1990), anthropological research and activism are qualitatively different efforts: Whereas anthropology draws legitimacy from scholarship to create knowledge, advocacy relies on moral legitimacy to apply knowledge. As such, ethnographic experiences may lead anthropologists to become activists, yet the rationale for such activism is not ethnographic. For Hastrup and her coauthors, advocacy involves an unacademic emotional rhetoric that is appropriate in some circumstances but risks jeopardizing anthropology's credibility. Similarly, Merry (2005) argues that research and activism are incommensurate in terms of epistemological principles: While human rights activists tell simple stories with clear villains, anthropologists elucidate more complex circumstances and define social injustice contextually. Academic research can thus inform activist endeavors.

How does the tension between activism and academic inquiry look in practice? After the lawsuit began, Luciano Lliuya ran into difficulties with some of his neighbors. As I discuss in chapter 2, people in his community had trouble understanding why he had filed a legal claim in Germany over glacial retreat in Peru. As Luciano Lliuya rose to fame, rumors abounded that he was making ill-gotten gains. Initially, Luciano Lliuya neglected to talk about this fact in media interviews. It did not fit the image of David versus Goliath: a symbolic representative of vulnerable Global South communities taking on the powerful corporation. I do explore Luciano Lliuya's troubles with his neighbors in my academic work; they point to a disconnect between global discourses and local experiences of climate change (Walker-Crawford 2023). Eventually, Luciano Lliuya became more open about his neighborly disputes, and tensions calmed over time.

In this book, I ask questions that may make activists feel uncomfortable. Activism is about doing things; anthropological inquiry asks why those things should be done and whether they achieve their stated aim or produce adverse impacts. Some anthropologists are open to combining academic work and advocacy. For Scheper-Hughes (1995), anthropologists should not pretend to act as rational objective observers; rather, anthropology should be morally engaged and committed to an ethic of care. In politically charged situations such as Apartheid South Africa, argues Scheper-Hughes, anthropologists have the duty to take a stand against violence and oppression. Ethnography creates a space for shared empathy and should serve as "a tool for critical reflection and for human liberation" (418). In a similar vein, Kirsch (2002) has argued that "activism is a logical extension of the commitment to reciprocity that underlies the practice of anthropology" (178). This stance led Kirsch to advocate on behalf of Indigenous groups in Papua New Guinea against a mining company that threatened local communities. In contexts marked by significant power imbalance, Kirsch argues, anthropologists should actively support subaltern groups.

My view is that asking difficult questions makes activism stronger. Politically, I am committed to working toward equitable solutions addressing climate change. At the same time, my advocacy is a productive site of ethnographic knowledge production. By participating in an international climate litigation claim, I gained unique insight into the dynamics of climate concerns as they emerged between Peru, Germany, and international discussion forums. My position gave me the opportunity to investigate the political and moral stakes of a climate litigation claim. My prolonged involvement and political commitment are indispensable to gaining this perspective. While ethnographic research has led some anthropologists to become activists, my participation in the climate justice advocacy led me to ask anthropological questions about why this cause is worthwhile. Activism involves making compromises; any strategy has its downsides. Asking difficult questions can help us understand these complexities so we can take a stronger stand in an imperfect world.

What This Book Does

This book tells the story of how Luciano Lliuya's lawsuit against RWE emerged, who it involved, and how it invoked moral relations at individual and universal scales through an appeal to neighborliness. I provide an otherwise unseen perspective on how climate change can be brought to court. This book shows why the lawsuit matters. With the very fact that it forced RWE's representatives to confront Luciano Lliuya in the courtroom and to see the melting glaciers with their own eyes during a court visit to Peru, it established a new kind of moral relationship that emerges uniquely out of anthropogenic climate change—between a major polluter and a human person who suffers the consequences.

In May 2025, as this book was going into production, the Upper State Court in Hamm issued its final verdict. While the case was dismissed on the grounds that the flood risk to the plaintiff's property was not sufficiently high to justify liability, the court affirmed a landmark principle: Major corporate emitters can, in principle, be held legally responsible for climate-related harms. This marked the first time a court anywhere in the world recognized the possibility of establishing a legal and moral link between large-scale polluters and those suffering the impacts of climate change. By affirming the principle of corporate climate liability, the ruling set a powerful legal precedent—one that will likely influence climate litigation worldwide and open new avenues for claims by affected communities.

This book offers a detailed ethnographic account of the case up to 2021. The latter stages (2022–2025) are discussed more briefly. While I remained involved to the end as both researcher and participant, an in-depth discussion of later developments—including the court's site visit to Peru in 2022 and final hearings in 2025—lies beyond the scope of this book. As an anthropological study, my approach is unusual in that it follows one person's legal claim around the world. I offer glimpses of how Peruvian communities and authorities dealt with glacial retreat, but this is not a community study of climate change in the Andes. Others have written extensively on that topic (Rasmussen 2015; Stensrud 2016a; Paerregaard 2020). Instead, I study the moral links that climate change creates across the planet, connecting people, corporations, and mountain beings. I do not answer the normative question of who should take responsibility for climate change, but show how climate activists, lawyers, and judges grapple with the issue.

It is common practice among anthropologists to change the names of our ethnographic interlocutors to protect their identity. That is not a viable option for the major characters in this story such as Luciano Lliuya and his lawyers, whose names have appeared in countless media reports. They deliberately acted as public figures advocating for climate justice. I changed the names of lesser-known people in this story, such as the glacial lake workers encountered in part 3, unless they specifically requested otherwise.

The book is divided into three parts. Part 1 shows how the claim emerged and outlines the legal questions at stake. The lawsuit used neighborhood law to link Luciano Lliuya with RWE, arguing that climate change makes us all potential neighbors. I examine the implications of this approach in relation to legal scholarship on climate change. Building on academic discussions about legal personhood and the rights of nature, I examine how the legal process brought together the plaintiff and defendant as legal persons but formally excluded Andean mountain beings, even though they played an important role for Luciano Lliuya.

Part 2 dives into the lawsuit's scientific and evidentiary arguments. Luciano Lliuya's legal team drew on climate change research to create a causal chain linking RWE to climate change impacts in Peru, turning the plaintiff and defendant into neighbors. I relate this causal argument to discussions in legal scholarship and in science and technology studies about the production of knowledge within the judicial framework. I unpack the arguments on both sides at each step in the chain: from RWE's coal-fired power plants to global warming, from global warming to glacial retreat, and from glacial retreat to flood risk affecting Luciano Lliuya's house in the Andes. I trace the production of facts from measurements taken at an Andean glacial lake to written evidence submitted to the court. I show how both the plaintiff's and defendant's scientific arguments are linked to broader ideas about who should take responsibility for climate change.

We go back to the Andes in part 3. The mountains may not count as neighbors in the German courtroom, but they play a central role in the local politics of glacial retreat. Many Andeans, including Luciano Lliuya, engage with mountains and lakes as living beings. They inhabit a sentient landscape that is undergoing rapid transformation. This involves numerous changes, such as glacial mass loss and flood risk, that can be measured with scientific tools. But, more than that, climate change has the potential to transform people's cosmological engagement with the landscape in ways that are difficult to ascertain. In their efforts to address flood risk and water scarcity caused by glacial retreat, Andeans employ both scientific engineering practices and appeals to powerful mountain beings. Sentient ecosystems are formally excluded from political processes, yet they pop up in unexpected places.

In the conclusion, I return to the main issue at stake in this book: Climate change expands moral relations around the planet. It makes everyone a potential neighbor—people, companies, sentient mountains, and countless others. I offer a methodological approach to studying cross-planetary moral relations. I discuss the limits of legal claim making in the contemporary context of climate politics and strategic litigation. Finally, I offer reflections on the role of anthropologists and other academics in addressing climate change, which ultimately threatens all our livelihoods around the world.

Part I

MAKING A CLIMATE CHANGE LAWSUIT

"This feels like we're at five thousand meters in the Andes," remarked Julio Luciano Shuan when it began to snow. On a fateful day in November 2015, Saúl Luciano Lliuya took a historic step by suing the German energy giant RWE to bring about climate justice. Fighting biting winds, we walked through cold streets up to the courthouse in the German city of Essen (figure PartI.1). It was Luciano Lliuya's first trip outside Peru, and he had a purpose: to hold RWE responsible for dramatic changes in his Andean environment. He had traveled with his father, who was remarkably fit at age seventy-five. I froze alongside them in my thick winter coat, but the two Peruvians sported only light jackets.

We trekked to the courthouse alongside a group of activists as a TV crew filmed our progress. I accompanied Luciano Lliuya as his guide and interpreter. Alongside us walked Roda Verheyen, Luciano Lliuya's lawyer. Now forty-three, she had spent most of her adult life fighting for the cause of climate justice. Representing Luciano Lliuya in the case against RWE was the pinnacle of her career so far.

FIGURE PARTI.1. Saúl Luciano Lliuya and Julio Luciano Shuan at the courthouse in Essen before filing the lawsuit against RWE, November 2015. (Photo: Germanwatch)

Luciano Lliuya and Verheyen entered the courthouse to file the lawsuit. I waited outside with a delegation from Germanwatch. We had worked closely with Verheyen to assemble legal and scientific arguments to hold RWE accountable for climatic risk in the Andes. With the lawsuit, Luciano Lliuya spearheaded the global cause for climate justice. Aside from reducing the risk of flooding that threatened Luciano Lliuya's house, the people at Germanwatch wanted the court to establish a precedent for holding major emitters accountable. This decision could have massive ramifications for global industry—past emissions could become an economic liability. Germanwatch's goal was and is to push energy producers toward adopting more sustainable business models to prevent even more devastating climate change.

Luciano Lliuya came to the claim from a different standpoint. He feels a profound connection with the mountains in his environment. As glaciers recede, the environment is under threat. He seeks to defend the mountains in Peru. After submitting the lawsuit, Luciano Lliuya spoke to journalists and TV cameras outside the courthouse: "I'm making this claim because the mountains in Peru are suffering. The glaciers are melting. We haven't caused this problem—it's big companies like RWE. Now they must take responsibility."

1

Glaciers Melt into the Courtroom

"I'll do it. I'll do the claim."

It was December 2014. Luciano Lliuya smiled and glanced around the table of his two-story adobe house with a decisive look on his face. I interpreted from Spanish to German for the three Germanwatch representatives, who were finishing their plates of guinea pig and potatoes with red chili sauce. Luciano Lliuya's family had prepared a special meal for an unlikely visit from foreign guests. In the family's native village, we sat at the table with Luciano Lliuya's father and several other family members.

"All right, then," said Christoph Bals, the head of Germanwatch, whose words I translated into Spanish for the Peruvians. He smiled and looked into Luciano Lliuya's eyes. "We're going to court!"

Earlier that week, we had begun discussions with Luciano Lliuya's father, Julio Luciano Shuan, about organizing a climate litigation claim against a major greenhouse gas emitter. Luciano Shuan owned a house in Huaraz that faces a major risk of flooding from the Palcacocha glacial lake.[1] According to scientific research, climate change is to blame: Glaciers are melting due to global warming, causing the lake's volume to rise.[2] Luciano Shuan had expressed interest in making a legal claim but told us over lunch in the village that he had divided his property among his seven children. To our surprise, Luciano Lliuya offered to make the claim himself. At thirty-four, he was the youngest of his siblings and the only son. Luciano Shuan looked to him fondly as we agreed to arrange a call with the lawyer in Germany.

Over their lifetimes, Luciano Shuan and Luciano Lliuya have witnessed dramatic changes in their mountain environment. In 1941, when Luciano Shuan was one year old, an outburst flood from Palcacocha devastated the city of Huaraz and killed thousands. The lake's natural moraine dam broke, washing a great mass of water, mud, and rock down into and through Huaraz (Wegner 2014). Living on higher ground in the upstream village, Luciano Shuan and his family evaded death. He traversed many of the region's glaciers when he worked as a mountain guide in the 1970s. Since then, the ice has receded dramatically. Stepping into his father's shoes, Luciano Lliuya became a mountain guide in the early 2000s. Both father and son share a deep concern for glacial retreat. They see glaciers as vital sources of water that enable agricultural livelihoods and as living beings that are suffering in uncertain times. They are not responsible for climate change; large industry and wealthy countries cause global warming and glacial melting. Yet Luciano Shuan and Luciano Lliuya lacked any means of taking action—until a mutual friend brought a small delegation of German climate activists into their village.

This chapter tells the story of how and why the lawsuit started. In a region scarred by colonial domination and environmental disaster, Luciano Lliuya and his compatriots struggle to confront the devastating impacts of climate change. Glacial retreat causes short-term flood risk and long-term water scarcity. It threatens people's cosmological foundations as they fear for the future of powerful Andean mountain beings. Climate change ruptures the fabric of Andean life. The lawsuit against RWE was one man's attempt to confront this existential crisis.

How It All Started

How did Luciano Lliuya come to collaborate with German climate change activists in an emblematic international legal claim? The NGO Germanwatch had spent two decades participating in UN Climate Summits to push for sustainable solutions to a global crisis. As politicians made little progress, activists sought new avenues of action, drawing on analysis from legal scholars about potential litigation strategies.[3] On other social and environmental issues, legal claims had advanced debates when politicians and industry failed to take action. It took dozens of lawsuits against tobacco companies over several decades to persuade the industry to acknowledge its responsibility for smoking-related health risks (Rabin 2001). In a similar vein, climate litigation could push for action against global warming. Since the early 2000s, activists and lawyers had discussed the possibility of making legal claims against major emitters. When I joined Germanwatch in 2014, I found ongoing conversations about possible strategies under German law, including claiming protective measures from a large emitter in German courts. Groups from countries such as Bangladesh had asked Germanwatch to assist in assessing possible legal pathways, but a claim was yet to emerge.

Leading up to the 2014 UN Climate Summit in Peru, Germanwatch employees took interest in the Cordillera Blanca in the northern Peruvian Andes. This mountain chain is emblematic for its vulnerability to climate change impacts. Numerous studies point to glacial retreat, flood risk, and long-term threats to water security (Hegglin and Huggel 2008; Vilímek et al. 2014). However, the people at Germanwatch were unfamiliar with the region and did not speak Spanish. As I had lived in Peru and knew the country well, I joined the team to find people in the Andes who shared our concerns about climate change.

I reached out to a friend, a Peruvian agricultural engineer who worked with small-scale farmers in animal husbandry projects. Traveling around Huaraz, he had struck up a discussion with Luciano Lliuya's father after a village assembly. At the time, Luciano Shuan held an official position of authority in the village. He was concerned about how climate change was affecting his community's future livelihood. After an introduction via my friend, we arranged to visit Luciano Shuan with a team from Germanwatch after the UN summit in Lima.

Following two weeks of intense UN negotiations, I embarked on an eight-hour bus ride to Huaraz with three Germanwatch representatives. They wanted to see how global warming was changing the Andean landscape and meet this farmer who shared their worries. Luciano Shuan and Luciano Lliuya met us in the city with their rickety Toyota van. It had seen better days but managed to carry them on the thirty-minute journey down an uneven dirt road from their village to Huaraz.

In our first discussions, my colleagues asked our interlocutors about their experience of climate change. I acted as an interpreter while father and son elaborated on their concerns: Glaciers were disappearing, and the community's future was uncertain. If the glaciers were gone, where would they find water to drink and irrigate their fields? And, in the short term, glacial retreat caused a risk of flooding as mountain lakes filled up to the brim. Luciano Shuan and his wife, Juliana Lliuya, had bought land in the Huaraz district of Nueva Florida in the 1980s. Luciano Lliuya spent part of his childhood living in a small adobe hut in Nueva Florida, where he had easier access to school. According to scientific studies, Nueva Florida could be swept away almost entirely if there were a flood (Somos-Valenzuela et al. 2016).

Located in the northern Peruvian Andes, the Cordillera Blanca is the country's largest glaciated area. Every year, it attracts tourists from around the world to trek its valleys and climb its immense peaks. The region has hosted numerous Peruvian and foreign scientists who have documented the impacts of climate change in the Peruvian Andes. The Cordillera Blanca has lost around 30 percent of its glacial cover since 1930 (Schauwecker et al. 2014), posing the combined risk of too much and too little water. Shrinking glaciers cause glacial lakes to rise in volume, leading to the danger of glacial lake outburst floods (GLOFs). In the longer term, glacial retreat causes water scarcity. In scientific terms, glaciers act as water storage devices and are particularly sensitive to climatic changes (Drenkhan et al. 2015). In the Cordillera Blanca, many villages depend on glacial meltwater for irrigation and household use. The ice usually grows during the rainy season and recedes during the dry season. Climate change has brought the glaciers out of balance, threatening people's livelihoods.

We wanted to visit Palcacocha and see the situation for ourselves. Luciano Lliuya offered to take us; as a mountain guide, he had often passed by the lake. Luciano Lliuya and his father picked us up at our hotel in Huaraz on an early morning. In their old van, we drove up dirt roads past their village to Cojup Valley. From there, we set off on a strenuous 11-kilometer hike up 700 altitude meters over uneven terrain to the lake, which sits at over 4,500 meters above sea level. The narrow green valley lay between dark, towering cliffs. We followed a narrow path past shrubs, small trees, and slender cows grazing on grassland. Alongside our walk ran a small river. "That water comes from Palcacocha," Luciano Lliuya remarked. "It's what we drink in Huaraz." Suffering from altitude sickness, two of my colleagues turned back halfway to the lake and returned to the van, where Luciano Shuan (age seventy-four) waited patiently. With Luciano Lliuya and Germanwatch director Christoph Bals, it took us six hours to reach the lake.

In the early afternoon, we approached the lake's natural moraine dam. It had a massive gap. "In 1941, the moraine dam broke, and the water flooded

FIGURE 1.1. Lake Palcacocha in October 2017. (Photo: Alexander Luna)

out [figure 1.1]. That's what destroyed Huaraz," explained Luciano Lliuya. By the moraine dam we found a group of stone huts. Light whiffs of smoke came out of a chimney. An old man approached us. Under a wide-brimmed hat, his face showed a wrinkled smile. After a brief exchange with Luciano Lliuya in Quechua, he spoke to us in Spanish.

"Welcome to Palcacocha!"

He introduced himself as Eduardo Díaz. As superintendent for the regional government's Palcacocha safety works, he received few visitors. In his late seventies at the time, Díaz comes from Luciano Lliuya's village and had worked in the mountains for most of his life. In the following years, Díaz would become a friend and important ethnographic interlocutor. At the time of our first meeting, he had worked at Palcacocha for over two years.

Díaz guided us through the broken moraine dam along ten large black plastic tubes. "We use these to pump water out of the lake," he explained as we made our way up. Finally, we reached a concrete dam. High up in the mountains under the burning sun, we struggled to take the last steps. At the top of the dam, we laid eyes on Palcacocha in all its magnificence. Blue water sparkled in the sunlight under shining white glaciers. Almost two kilometers across and five hundred meters wide, it felt massive.

As we approached the water's surface, a distant crash broke through the silent wind. The glacier ice was cracking. Noticing our alarm, Díaz told us not to worry: "It's just a small avalanche. That happens all the time." Far away, on the glacier, I spotted a flurry of falling snow. "You see, this one didn't even reach the lake." This was a regular occurrence, I came to learn. But, according to scientists studying glacial hazard in the area, a large avalanche falling into the lake could lead to waves overtopping Palcacocha's concrete dam, potentially causing a deadly flood in the valley below (Somos-Valenzuela et al. 2016).

After we had spent an hour by the lake, it was time to head down. Walking back through the valley with my colleague from Germanwatch trailing behind, I chatted with Luciano Lliuya about his experience with climate change. As a mountain guide, he climbed the melting glaciers year after year. "When I come back to some glaciers this year, they will have retreated by twenty or thirty meters. It's really extreme." Glaciers provide the region with water, and if there was no more water, what would be left of life? With troubled eyes, he gazed down the distant valley. For Luciano Lliuya, climate change is an existential threat.

Climate Change Alters the Fabric of Andean Life

The Cordillera Blanca is a comparatively young mountain range that started forming around thirteen million years ago. Its peaks are distinctly ragged. Glaciers are attached directly to summits and rock walls, making them unstable and prone to breaking off. The cordillera contains numerous glacial lakes that threaten to spill over their banks and flood the valleys below. Its towns lie in exceptionally close proximity to large glaciers (Bode 1989, 4–6).

For the past six hundred years, the Cordillera Blanca's original inhabitants have faced successive domination by Incas, Spaniards, and European-descended rulers of the Peruvian republic (Bode 1989, 8). In the fifteenth century, the Inca Empire expanded from its capital of Cusco in the southern Andes and took control of the Cordillera Blanca. The Incas forced people in the region to speak Quechua and eradicated other languages. Contemporary Quechua in the Cordillera Blanca notably differs from the Cusco variety in terms of vocabulary and grammar, pointing to the influence of now-extinct languages. In the sixteenth century, Spanish colonizers defeated the Inca Empire and implemented a system of forced labor. After Peruvian independence in 1821, a system of large landholdings (*haciendas*) kept rural Andeans working in serf-like conditions for the benefit of the country's elites, most of whom were descendants of Spanish colonizers (Bode 1989, 7–8). A 1969 reform moved land ownership into the hands of small-scale farmers (Poole

2004), yet marginalization of Quechua-speaking villagers is ongoing. They face high rates of poverty and unequal political representation.

Today, there is a persistent divide between city and countryside: Urban residents tend to be wealthier and to speak Spanish, while rural areas are poorer and most speak Quechua. Discrimination is not uncommon; many members of the urban elites regard rural people as less developed (Rasmussen 2015, 28). In 2019, poverty in rural areas across the region was over 36 percent, compared to 6 percent in the cities (IPE 2019). Illiteracy rates are much higher in rural areas, especially among older women. In 2017, around 20 percent of rural households had no running water or electricity (INEI 2018). While figures for poverty, education, and basic infrastructure have significantly improved in the past twenty years, there is still a stark urban-rural divide. In the villages around Huaraz, many feel abandoned by the state authorities. Numerous roads are falling apart, and people face irregular electricity outages. Government authorities appear unable or unwilling to address people's needs.

Most people in rural areas farm small plots of land, growing potatoes, corn, wheat, and other crops that can survive at high altitudes. Many own cows, sheep, and guinea pigs (the last a regional delicacy). Families usually eat the bulk of what they produce and sell the excess at the market in Huaraz. Rural life revolves around the agricultural calendar; people of all ages help out in the fields. Agriculture alone does not provide a large income. Many, especially men, look for work in the city and farther afield. They work in mining, construction, or whatever else is available. Some women are domestic workers for urban families.

The most significant sector for villagers in the region is tourism. The Cordillera Blanca is a popular destination for trekking and mountain climbing. Every year, tens of thousands of tourists from other parts of Peru and around the world visit the region. Villagers know the mountains well and make good guides. The tourist season, which lasts from around April to September, when the weather is mostly dry, provides a lucrative source of income. Many—almost exclusively men—have obtained certification from the Peruvian Association of Mountain Guides. Others work as cooks, porters, or mule drivers. In most rural families I have encountered, at least some of the men work in tourism. Many aspire to open their own tourist agency to capture the services of wealthy foreigners who want to climb the region's magnificent peaks.

After the 1941 Palcacocha flood, which destroyed much of the city, government officials in Huaraz set up a dedicated authority to address glacial lake hazard.[4] Between the 1950s and the 1970s, the Glacier Authority implemented numerous safety projects at glacial lakes in the Cordillera Blanca, building dams and drainage systems to reduce flood hazard. In 1974, authorities completed two concrete dams and a drainage canal at Palcacocha. While glaciers have retreated at unprecedented rates since then, Peruvian authorities' capacity to address glacial lake hazard was significantly reduced following cutbacks and privatizations in the 1990s and political decentralization since the 2000s (Carey 2010). Today, the Glacier Authority exists only as a monitoring agency. The Ancash Regional Government is formally responsible for directing glacial lake safety efforts in cooperation with national and local public authorities. Following this institutional fragmentation along with governmental instability and widespread corruption allegations at the regional level, authorities have been slow to respond to increased flood hazard at Palcacocha and other lakes. Part 3 of this book explores these dynamics in more depth, highlighting how international discussions about climate change renewed older concerns about flood hazard in the Cordillera Blanca.

On May 31, 1970, Huaraz was hit by a magnitude 7.9 earthquake that destroyed almost the entire city. Most structures were made of adobe bricks and could not withstand the Earth's violent force. With a population of around 65,000 at the time, 20,000 residents of Huaraz lost their lives (Bode 1989, 30). North of Huaraz, the earthquake caused a massive avalanche at Huascarán, the highest mountain in Peru. It buried the town of Yungay under a mass of ice and debris, leaving around 15,000 dead (Carey 2010, 130).

After the earthquake, geologists declared Huaraz unsafe, and authorities planned to relocate the city. But residents resisted, not wanting to leave their homes (Bode 1989, 79). As Huaraz was gradually rebuilt in the 1970s, Quechua-speaking farmers from nearby villages in the Cordillera Blanca settled in the city. North of Huaraz lay a patch of land devastated by the 1941 flood that nobody had sought to resettle. Strewn with boulders, the land was cheap and provided easy access to education and work opportunities in Huaraz. With no public transportation to the villages at the time, farmers had to walk for hours to reach the city. Villagers built small mud-brick houses there, and the area eventually emerged as the district called Nueva Florida. Most retained their homes and agricultural plots in the village, commuting between city and countryside.

According to early settlers who arrived in Nueva Florida in the mid-1970s, local authorities initially sought to prevent construction in the area, which they had considered unsafe since the 1941 flood. In addition to Palcacocha, it sits below Lakes Cuchillacocha and Tullpacocha. The settlers refused to give up on their newfound opportunity to live near the city, however, and eventually officials relented. By the 1980s, the authorities had largely left Nueva Florida to its own devices (Huggel et al. 2020). Among the settlers in the mid-1980s were Luciano Lliuya's parents. This property later became the subject of Luciano Lliuya's lawsuit against RWE.

As the district grew, city authorities eventually installed roads and electricity. The district expanded after a boom in multinational mining brought laborers from other parts of Peru to Huaraz. Over time, property values in Nueva Florida increased astronomically. Whereas a square meter had brought only a few US cents in the 1980s and 190s, prices ranged between US$300 and US$400 per square meter in 2025. The early settlers had made a fruitful investment. Meanwhile, Palcacocha swelled because of glacial retreat and by 2009 reached an even greater volume than before the 1941 disaster. If that event repeats itself, the flood wave would wash through Nueva Florida before inundating lower areas in Huaraz.

Born in 1980, Luciano Lliuya came of age in a changing landscape. He grew up in a village near Huaraz and went to school in the city. Quechua is his first language, as it is for most villagers in the area, and he learned Spanish at school. Like other children from Quechua-speaking families, he faced discrimination: Teachers made fun of him when he struggled to answer questions in Spanish. Since his childhood, Luciano Lliuya enjoyed roaming the mountainous landscape. Growing up, his father told him stories about guiding foreign tourists to the region's peaks and working at glacial lake infrastructure projects in the 1970s. After finishing school, Luciano Lliuya studied at the Peruvian Association of Mountain Guides, a prestigious local institution that offers a grueling three-year course to become a licensed mountain-climbing guide. The qualification offered Luciano Lliuya and his family a lucrative income during the yearly climbing season. While he and his family continue to farm the fields around their house, thereby sustaining themselves and selling excess produce at local markets, Luciano Lliuya's earnings as a mountain guide enabled them to build a house, buy a used Toyota station wagon, and send their oldest child to university. Today, Luciano Lliuya continues to live in his native village with his wife and two children, a stone's throw from his parents' house where he was born. The family frequently commutes between the village and their house in the Nueva Florida district of Huaraz. Luciano Lliuya's son and nephews live in the house permanently and attend university in Huaraz.

People in the Cordillera Blanca face new difficulties in the context of climate change. As glaciers retreat at an accelerated pace, many people, particularly in rural areas, are concerned about future water availability. Scientific research confirms farmers' worries (Jurt et al. 2015). As in other parts of the Peruvian Andes, villagers around Huaraz have come into contact with scientific discourses on climate change through government authorities, NGOs, and discussions with foreign tourists. These discourses inform villagers' observations of glacial retreat and water scarcity (Stensrud 2019b). Climate change is rapidly and irreversibly transforming people's experience of their environment (Stensrud 2016a). Locals point to a whole range of changes: the rain feels colder, the sun feels hotter, they have difficulty sowing as seasonal rain begins irregularly, frost damages crops and kills animals, pastures are disappearing, and children are falling ill more often.[5] People express these worries increasingly in terms of global climate change.

More than altering the physical environment, climate change can transform people's cosmological engagement with the landscape. *Cosmology* refers to people's philosophical and spiritual ideas about the universe and humanity's place within it. Many Andeans, including Luciano Lliuya, engage with mountains and lakes as living beings. They inhabit a sentient landscape that is rapidly transforming. While scientific tools can measure glacial retreat and flood risk, they cannot evaluate how climate change is making the mountains feel. Anthropological research has documented Andean belief systems and rituals, especially in the southern Andes, that involve relations of reciprocity between people and nonhuman beings in the landscape.[6] Further north in the Cordillera Blanca, such practices are less common. During my fieldwork in Peru, I met only a handful of people who openly discussed making ritual offerings to the environment. Lake workers at Palcacocha made offerings to the mountains to appease them and prevent a deadly flood. Luciano Lliuya and many other mountain guides carry a bag of coca on their tours that they offer to the landscape to ensure a safe journey. I found that people in rural areas rarely talk about engaging with the environment's sentient force, at least not with a foreign researcher. Yet the landscape seems to have an important cosmological significance. As I gained more trust with people, they told me about their encounters with sentient mountains and lakes in their dreams and while exploring high altitudes. When I told Luciano Lliuya at the beginning of my fieldwork about anthropologists' interest in Andean Earth Beings, he suggested we investigate the topic together. He knows their power and feels their suffering in times of glacial retreat, but is unsure about what they are. This book is, in part, the result of our joint exploration of the nature of the Andean mountains and how climate change influences them.

Governing Glacial Retreat

Government authorities in Huaraz have struggled to respond to the challenges posed by climate change. In recent decades, they have devoted significant attention to glacial lake outburst flood risk, seen as a threat to human life and regional economic development. In rural areas, many do not consider flooding as the most important issue. Given widespread mistrust of urban government officials, some even argue that there is no danger. However, villagers express grave concern about water scarcity in the context of accelerated glacier retreat, fearing a threat to their agricultural livelihoods. While those who are familiar with scientific discourses relate the problem to climate change, many blame the government for its failure to establish secure water infrastructure and to implement other measures to support farmers. People's perception of risk and danger is shaped by how they engage with the landscape and whom they trust to provide reliable information.

Since discovering in 2009 that Palcacocha had grown to a dangerous level, government authorities in Huaraz have made plans to implement flood risk reduction measures. After long delays, they set up an early-warning system in 2021. According to authorities, a flood would take around forty-five minutes to reach Huaraz. Officials plan to build a new dam and drainage system at the lake. In a region plagued by accusations of corruption and political instability, these projects have progressed slowly while climate-related risk increases steadily.

As an interim measure, authorities directed the installation of ten plastic siphons that continuously pump water out of Palcacocha. This reduced the water level by several meters, but the risk is still high. In the meantime, a small team of men, most of whom live in nearby villages, monitor the lake day and night. Isolated at a high altitude, they keep in contact with city authorities via a two-way radio and spotty internet connection. The lake workers perform maintenance on the siphons, ensuring the provisional flood risk infrastructure continues to function. They witness climate change on the front line, sleeping in a small shack above Palcacocha with a glorious view of the melting glaciers. Part 3 of this book follows their struggle to contain the lake's violent forces.

Luciano Lliuya Takes a Leap

Luciano Lliuya first met Roda Verheyen, the lead lawyer on the case, in a series of video calls following the Germanwatch team's visit to Peru in 2014. They discussed the possibility of taking legal action and considered various options, including lawsuits in Peruvian, German, and international courts. As Luciano Lliuya's preferred option was to address a major polluter, the team settled on a civil law claim against RWE. Although the chances of legal victory were low, Verheyen explained, if they won, the case would make a massive difference in the world.

Luciano Lliuya agreed to join us and an international network of activists in a precedent-setting claim for climate justice. This commitment marked a radical change of course for a man who had lived his life in the global periphery. It took him to German courtrooms and UN Climate Summits where he captured the passions of a burgeoning transnational climate justice movement. A man who had once been nervous about speaking at village assemblies went on to address thousands of people at major climate marches. He would give countless interviews to the world press (figure 1.2). His lawsuit achieved greater success than Luciano Lliuya had ever imagined. I accompanied him throughout this process as an interpreter, confidant, fellow activist, and ethnographer.

At this point you might ask: Who is the author of the claim? Is it Luciano Lliuya, Germanwatch, or the lawyers? Luciano Lliuya did not participate in writing legal documents or developing legal strategy; the lawyers handled this with support from Germanwatch staff. Luciano Lliuya's German collaborators sought to enact legal and political change; his most significant motivation was to support the mountains that are losing their glacial cover year after year as he watches. Luciano Lliuya and his interlocutors forged a pragmatic alliance and embraced a shared cause. As the lawsuit gained public attention in Germany and beyond, its collaborative nature faded into the background. Media profiles often focused on Luciano Lliuya as a lone man struggling for justice. Luciano Lliuya fit into a useful narrative mold: a historically subjugated subject from the Global South—possibly Indigenous, though he never referred to himself in those terms—taking on a powerful multinational corporation. This story lent the lawsuit emotive strength; in fact, it inspired me to write this book.

As this book goes to press in 2026, the case has reached its conclusion. Although the court ultimately dismissed the claim on factual grounds, it affirmed a landmark legal principle: Major corporate emitters can, in principle, be held liable for climate-related harms. This breakthrough ruling sets an important precedent in transnational climate litigation that will likely shape legal strategies, political discourse, and future claims around the world. The lawsuit came to define much of Luciano Lliuya's life for a decade. Several documentary film projects about him are currently in production. The claim emerged collaboratively

FIGURE 1.2. Luciano Lliuya on the cover of the Peruvian daily newspaper *La República*, March 30, 2018. Translated, the tagline reads: "Peruvian wins first round against German corporation."

as a shared project. Luciano Lliuya himself expresses this most succinctly: He has always referred to it as "our claim."

"Who are 'we'?" I once asked him after he used that formulation in a press interview.

"I say 'we' because it's not just me in this claim," he explained. "I have friends who are helping me with this claim, so they are also part of it."

In a conversation several years later, Luciano Lliuya reflected on his fateful decision to participate in the lawsuit and explained that he felt a responsibility to take action concerning glacial retreat: "It's something that had to be done." Referring to his own feelings, Luciano Lliuya usually speaks in the second person: "If you have the opportunity to do it, you should. Were there risks? Of course. But you just felt like you had to do it."

2

David and Goliath in the Courtroom

"A case like this would not exist in the industrialized world," explained Judge Rolf Meyer. He was heading a panel of three judges in the Upper State Court in Hamm, Germany, which was hearing the historic legal case *Luciano Lliuya v.* RWE.

"In a place like Germany," Judge Meyer said, "this problem would be solved immediately by building a dam or implementing other necessary measures." He glanced around the courtroom, a large, bright hall with ceiling-high windows facing a park. To his left sat Luciano Lliuya with two lawyers who were delighted that the case was achieving unexpected success. Luciano Lliuya followed the proceedings in Spanish via a court-appointed interpreter (figure 2.1). Long discussions of legal technicalities had left him confused, yet he was happy that the judges were taking the case seriously. Across from them sat RWE's legal team: five middle-aged men in dark suits, visibly annoyed by the judge's statements. Looking on was an audience of about seventy-five people, primarily climate activists and journalists. It was November 2017, two years after Luciano Lliuya initially filed the lawsuit against RWE.

FIGURE 2.1. Luciano Lliuya at the Upper State Court in Hamm, November 2017. From left to right: Roda Verheyen, Luciano Lliuya, the court-appointed interpreter, and the author. (Photo: Alexander Luna)

The case drew on legal norms that people usually invoke to seek relief from neighbors for damage or potential harm to their property. This strategy can involve a nuisance relating to environmental pollution if claimants can prove their neighbor's responsibility. In their arguments, Luciano Lliuya's lawyers expanded the idea of neighborliness to encompass relations across the planet: As climate change connects RWE and Luciano Lliuya, it makes them neighbors. The legal approach defines as neighbors those who are able to act on each other. Neighborliness emerges out of concrete claims that construct ethically charged relations between legally defined entities such as humans and corporations.

The lawsuit emerged when activists became disillusioned with political negotiations on climate change at the UN and developed legal mechanisms to address the problem at another level. The field of climate litigation has exploded since Luciano Lliuya's lawsuit began. It remains one of the most emblematic cases around the world. This chapter tells the story of how Luciano Lliuya came to confront RWE in the German courtroom by drawing on long-standing legal provisions about how neighbors should treat each other. I compare the legal approach to Luciano Lliuya's experience of neighborly relations in his village as he confronted misunderstanding and critique. The claim is based on a fundamental ambiguity: It invokes moral relations at an individual level, between Luciano Lliuya and RWE, and on a global scale, between all emitters and all those impacted. Using the concept of neighborliness as an analytical starting point, I unpack the contradictions and productive tensions that brought Luciano Lliuya to the forefront of climate justice activism.

Bringing Climate Change to the Courts

Since Luciano Lliuya took RWE to court in 2015, the number of climate change lawsuits has grown dramatically worldwide. One database identifies almost three thousand cases as of 2025 (Setzer and Higham 2025).[1] Around one-quarter have been filed since 2020. With his lawsuit, Luciano Lliuya joined a growing movement of lawyers and activists who are moving climate change from political discussions to the sphere of law. But why bring climate change to the courts?

For the people at Germanwatch, the lawsuit arose from two decades of efforts in fighting climate change. Klaus Milke, cofounder and former chair of the NGO's board of directors, had participated in the UN climate negotiation process since the first summit in 1995. For years, he watched governments flounder in their response to climate change. In 1997, he was in Kyoto to see states find consensus on the first major climate change agreement. The Kyoto Protocol set binding targets for Global North countries to reduce greenhouse gas emissions. He was disappointed when the United States refused to ratify the agreement and it did not produce the desired effects. In 2009, Milke was in the room when governments failed to achieve a follow-up agreement in Copenhagen. Member states finally approved the historic Paris Agreement in 2015, yet greenhouse gas emissions have continued to rise. Meanwhile, many countries struggle to cope with devastating climate change impacts. Even with the Paris Agreement, governments are not yet on a path that will likely lead to significant emissions reductions and prevent major climatic catastrophes (Boehm et al. 2022).

Luciano Lliuya's lawyer Roda Verheyen knew Klaus Milke from the early UN Climate Summits in the 1990s where she participated as a campaigner for Friends of the Earth and later with the German government delegation. In the early 2000s, she wrote her PhD thesis on legal liability for climate change (Verheyen 2005). As Verheyen and Milke became frustrated with the slow response to climate change at the UN level, they gave increasing thought to the possibility of legal action against governments and major emitters.

If politicians failed to act, ordinary citizens could seek protection against climate change impacts based on existing legal provisions. Climate change litigation often emerges in response to a perceived institutional failure to address climate change (Fisher 2013). Given unsatisfactory political action, some have argued that it can serve as a tool to fill regulatory gaps and as a catalyst for policy change (Peel and Osofsky 2015).

Climate change is relatively new as a legal phenomenon. As a broad, multiscalar issue in terms of its causes and impacts, it raises complex legal questions: Who can make claims? What should accountability mean in the context of climate change? Which issues are legally relevant? Climate litigation opens the door to addressing climate change and formulating a response within the existing legal order (Fisher 2013). Judges must determine whether issues raised by climate change can be resolved through the legal process (Burgers 2020). Legal systems place tight bounds on who can make claims and what kinds of claims can be made. Coming into the legal sphere, climate activists must distill a broad set of concerns to those that can be addressed within the judicial framework. While climate change affects us all, the legal system individualizes the problem as suits may be brought only by individual legal persons, be they humans or corporations.

What does climate change litigation hope to achieve? In public narratives surrounding the claim against RWE and other cases, litigation has been portrayed within the broader framework of an effort for "climate justice." Activists use this term to inject moral, political, and legal dimensions into global discussions of climate change. In their terms, seeking climate justice means promoting political and legal processes for holding major greenhouse gas emitters accountable for climate change and for supporting those who face its worst impacts. Climate lawsuits often seek to influence law, policy, and corporate behavior as well as shift public attention (Wonneberger and Vliegenthart 2021). In its proponents' eyes, climate litigation can shape public, political, and corporate incentives for action in relation to climate change (Peel and Osofsky 2015).

The claims themselves are usually particular, revolving around how specific actions affect individual actors. Strictly speaking, the lawsuit against RWE concerned the alleged violation of the plaintiff's right to enjoy his property. Yet Luciano Lliuya and his supporters hoped to set a legal precedent that other people affected by climate change could use to hold greenhouse gas emitters liable—that is, to make polluters pay.[2] Legal scholars have highlighted the broader significance of Luciano Lliuya's claim and its potential "far-reaching impacts" (Ganguly et al. 2018, 862). The threat of legal liability could also pressure politicians to take definitive action, argue climate activists (Frank et al. 2019). Polluting companies may urge policy makers to address climate change and thus avoid future litigation. This is particularly relevant for negotiations over Loss and Damage in the UN climate policy process: Litigation gives credence to vulnerable countries' demands that wealthy countries should help pay for dealing with impacts such as sea level rise and extreme weather events (Toussaint 2021).

Litigation may also impact financial markets, affecting costs and business risks for large corporate emitters. According to one study, companies' stock value decreased when climate cases were filed against them or surpassed major legal hurdles (Sato et al. 2024). Corporations are facing increasing numbers of climate lawsuits. They face potential liability for mismanaging climate risks, misleading investors about climate change effects on business, and failing to comply with legal reporting requirements. With this liability, climate change potentially extends from an externalized public safety risk to an internalized corporate risk (Ganguly et al. 2018). Activists hope that litigation will motivate corporations to move away from fossil fuels and promote investment in renewable energy.

Climate litigation can also focus greater media and public attention on climate change. Court judgments in climate change cases can force governments and private corporations to take climate change seriously (Preston 2021a, 2021b). Legal cases and the surrounding publicity can influence social norms and values relating to climate change (Peel and Osofsky 2015, 49). *Luciano Lliuya v. RWE* has already led to broad public discussions in Germany and elsewhere about responsibility and climate change. Major international media outlets profiled Luciano Lliuya, including the *New York Times*, *The Guardian*, and the *Financial Times*. Courts help legitimize concerns about climate change when they authoritatively state relevant facts in the course of adjudication, regardless of the final outcome (Fisher et al. 2017). Even unsuccessful cases can contribute to social change by helping to shift people's attitudes (Ganguly et al. 2018).

In the long run, Verheyen (2005) argued that the most effective social response to climate change damage is certainly not for all small-scale farmers or property owners like Luciano Lliuya to take major companies to court. High legal costs are, among other factors, a barrier for most potential litigants. Rather, politicians should establish mechanisms to help people like Luciano Lliuya cope with climate change impacts. To achieve climate justice, argue activists, the main contributors to climate change—both companies and countries—should help pay for adaptation measures and compensate those harmed for damages they suffer (Boom 2016). In using the law to put pressure on political institutions, climate litigation is a strategic effort to produce regulatory and social change.

There are two broad types of climate litigation: public and private cases. Public litigation involves cases against governments demanding more ambitious climate action. These claims have been the most successful so far. The first major ruling was made in 2015 when a Dutch court found that the government had taken insufficient action to reduce greenhouse gas emissions, endangering the human rights of its citizens. This legal strategy has since been replicated in numerous countries. The verdict was confirmed in 2019 by the Dutch Supreme Court, and in 2021 the German Constitutional Court found that climate change affects the rights of future generations. Both decisions forced governments to increase emissions reduction efforts. Similar cases have succeeded in other countries, such as France and Ireland, and numerous claims are currently under way. In the United States, a Montana court ruled in 2023 in *Held v. Montana* that the state had violated young people's constitutional rights by promoting fossil fuel development without considering climate impacts.

Private litigation targets corporations. It seeks action from private entities, often in relation to climatic risks and damages. Most cases are either forward-looking or backward-looking. Forward-looking cases mirror the claims against governments, demanding that corporations cut their future emissions to limit dangerous climate change in line with the Paris Agreement. The first major initial success was a suit by the NGO Milieudefensie against Royal Dutch Shell in the Netherlands. In 2021, the District Court of The Hague ruled that Shell must reduce its global greenhouse gas emissions as measured in 2019 by 45 percent by 2030. This verdict sent shock waves through the fossil fuel industry, and the legal strategy has been replicated in claims against the French oil giant Total and the German car industry. In 2024, the verdict was overturned on appeal in what amounted to a pyrrhic victory for Shell: The Hague Court of Appeal struck down the specific emissions target but affirmed that Shell had a legal duty to reduce its emissions.[3] Although Shell avoided a quantified obligation, the ruling nonetheless established a critical legal precedent. The case was on final appeal at the Dutch Supreme Court as of 2025.

Other corporate climate lawsuits, including *Luciano Lliuya v. RWE*, are backward-looking: They argue that corporations should accept liability for their contribution to existing climate change impacts and pay compensation. None of these claims have led to successful verdicts as of 2025. The

RWE claim advanced the furthest worldwide. In the United States, around three dozen cases have been filed by city and state governments against major fossil fuel companies. Plaintiffs include California, New York City, and Washington, DC. They argue that companies produced carbon emissions despite knowing about the dangers of global warming. The claims demand financial compensation and support in dealing with the effects of climate change. Many of these cases have been mired for several years in procedural disputes over whether they should be heard in federal or state court and are pending as of 2025. A more recent legal strategy combines the forward- and backward-looking approaches: Threatened by sea level rise, Indonesians on the island of Pari initiated a lawsuit in 2022 against the Swiss cement company Holcim. They demand compensation for existing impacts and demand that the company reduce its future emissions to prevent further damage. As climate science rapidly advances, the evidentiary base for compensation claims is improving. Cases are emerging against banks that finance fossil fuel projects and against company directors to hold them personally responsible for failing to address climate change. Inspired by initial successes, lawyers are devising creative strategies for holding corporations to account.

"We Live at the Bottom of a Sea of Air"

The legal claim that Luciano Lliuya submitted to the court in November 2015 was a thirty-nine-page document. The first half drew on scientific literature to argue that RWE contributed to glacial flood risk affecting Luciano Lliuya's property. The latter half of the lawsuit justified the claim in terms of German law. Roda Verheyen wrote the document in cooperation with Germanwatch employees, including myself. With support from legal colleagues, Verheyen developed and formulated the lawsuit's legal argumentation. At Germanwatch, we read countless scientific papers on climatic processes, glaciology, and glacial lake outburst floods. We collected Peruvian government reports and media articles about the situation at Palcacocha. After compiling this information according to legal requirements, we helped Verheyen write the legal text that she and Luciano Lliuya later filed at the courthouse. Along with the thirty-nine-page complaint, they submitted a much larger stack of attachments, including scientific studies, Peruvian government documents, and the deed to Luciano Lliuya's property.

In legal terms, Luciano Lliuya's lawyers argued that the German energy producer RWE has caused a nuisance to Luciano Lliuya's property in Huaraz. The plaintiff and defendant were configured as neighbors. Climate litigation cases can prompt judges to reconsider fundamental legal categories (Kysar 2011). Faced with the challenge of climate change, this lawsuit urged German courts to expand their understanding of legal liability and what it means to be a neighbor.

According to the lawsuit,[4] emissions from RWE's coal-fired power plants located in Europe contributed to the concentration of CO_2 and other greenhouse gases in the world's atmosphere. These gases insulate the planet by retaining a larger portion of solar energy, thereby producing the greenhouse effect and global warming. This insulation has led to glacial retreat around the world. In Peru, the lawsuit recounts, Andean glaciers have melted at a particularly fast rate. As a result, glacial lakes such as Palcacocha have grown in volume, increasing the risk of flooding. Palcacocha grew dramatically from around 0.5 million cubic meters in 1974 to over 17 million cubic meters in 2009. Several floods have occurred at Palcacocha and other Cordillera Blanca lakes in recent decades. In the city of Huaraz, Luciano Lliuya's house sits in the path of a potential outburst flood.

Citing flood models devised by University of Texas scientists, Luciano Lliuya's lawyers stated that a flood from Palcacocha threatened to destroy his house.[5] They argued that RWE and other greenhouse gas emitters were partially responsible for this flood risk. To remove the threat to Luciano Lliuya's property and the city of Huaraz, the regional government planned to build a new dam and drainage system at Palcacocha. Authorities valued this project at around US$4 million. The lawsuit sought from RWE not the entire sum but only a partial payment in accordance with its contribution to greenhouse gas emissions. According to the Carbon Majors Report, which quantified historical emissions and linked them to individual companies (Heede 2014a), RWE was responsible for 0.47 percent of industrial emissions between 1751 and 2010. Following this logic, the lawsuit demanded that the company pay 0.47 percent of US$4 million, or US$18,800, to the regional government toward the Palcacocha safety project. The lawsuit made an alternative secondary claim that RWE reimburse Luciano Lliuya for the cost of strengthening his house against flooding, but this was to be applied only if the first claim—that RWE contribute toward the government project—failed on legal grounds.

Courts face a significant challenge when they apply old legal doctrines, formulated before contemporary concerns with climate change, to the complex and multiscalar problem of global warming (Osofsky 2007a). Luciano Lliuya's lawyers addressed these challenges by reinterpreting an old legal norm in order to draw the causal link between RWE's emissions in Europe and climate change impacts in Peru. The lawsuit asserted a relation between Luciano Lliuya and RWE in accordance with Section 1004 of the German Civil Code:

> *Claim for removal and injunction*
> If the ownership is interfered with by means other than removal or retention of possession, the owner may require the disturber to remove the interference.[6]

German lawmakers passed this law in 1900 as a general nuisance provision. Lawyers have typically used it to resolve neighborhood conflicts. If one (legally constituted) person causes harm or risk of harm to another person's property, the latter person can sue the former, citing Section 1004, and demand that they remove the interference. This is a key provision in German law regulating relations among neighbors.

Verheyen drew a simple analogy to explain the legal approach:

> Imagine if your neighbor has a wall that divides their property from yours. The wall is old, and the bricks are loose, so you're afraid it could fall onto your property and damage your house. If that happened, you could sue your neighbor for damages. But you would rather not wait. You don't want to live with the uncertainty—when will the wall fall over? So you sue your neighbor over the hazard, citing Section 1004. You force them to remove the problem. In this case, make them fix the wall. In Luciano Lliuya's case, remove the flood hazard.

In legal terms, she argued that Luciano Lliuya and RWE are neighbors. This argument built on other cases that have addressed local environmental harms such as noise and smell pollution via Section 1004. In subsequent legal arguments, the lawyers cited historical German jurisprudence that defines neighborly relations in broad terms;[7] accordingly, the neighborhood is as large as potentially harmful effects can reach.[8] Following this legal logic, any greenhouse gas emitter is a potential neighbor to someone who faces climate change impacts. The "neighborhood" encompasses the entire planet. By invoking a neighborly relation, one could apply Section 1004 in a German court for claims relating to harm caused anywhere in the world if the defendant is based in Germany.

Defendant RWE's lawyers disagreed with this construction of Section 1004. In its first response to the lawsuit, the company argued that "German civil law provides no basis for liability in cases of potential interference by 'all against all' due to global climate change."[9] According to this interpretation of the law, Section 1004 is not applicable to climate change cases. Even if it were, RWE questioned the causal link between its emissions and specific climate change impacts in Peru. In their legal briefs and oral arguments, RWE's lawyers did not deny the existence of anthropogenic climate change, or that the company's emissions have contributed to global warming. Rather, they argued that the processes of climate change are too complex to draw a causal claim in terms of legal liability between an emitter and a specific impact. They disagreed with the articulation of a neighborly link between Luciano Lliuya and RWE framed in terms of an individualistic legal relationship.

German lawmakers drafted Section 1004 in the late nineteenth century, many decades before greenhouse gases and climate change became major public issues. Nevertheless, given early industrial pollution, environmental concerns may have been on their minds. Verheyen drew on an old legal text explaining the motives for the civil statute which asserted that neighbors are not only those who can see or hear each other.

At the court hearing in November 2017, Judge Meyer went into great detail summarizing Luciano Lliuya's legal claims. Then he quoted the official commentary accompanying the law that Luciano Lliuya's lawyers had cited in their submissions:

> Some types of effects cannot be kept within specific boundaries. We live at the bottom of a sea of air. This circumstance necessarily means that human action extends into the distance. . . . If the permission or prohibition of such an immission is to be determined,[10] one must not only consider the relationship of neighbor to neighbor; rather, the scope of the owner's right can be made to bear on all people. . . . Someone who causes or spreads imponderabilia must know that these go their own way. Their propagation across the border can be attributed to them as a consequence of their action. (Mugdan 1899, 146)[11]

"Prophetic phrases," Judge Meyer said. In a lower court in Essen, judges had dismissed Verheyen's argument, ruling that Section 1004 could not be applied to create a neighborly relation in the context of climate change. In the appeals court in Hamm, the justices disagreed with the previous ruling—much to the surprise, it appeared, of RWE's lawyers. In the German legal system, appeals courts can take new evidence and effectively act as trial courts in complex civil cases.[12] On the plaintiff's side, Verheyen became ever more excited. As a layperson, Luciano Lliuya did not follow all the technical legal details transmitted to him by the court-appointed interpreter. Nevertheless, he understood that the hearing was going well. Verheyen gave him looks of joy and squeezed his hand. She and her colleagues had found success in applying an old law to climate change, drawing on a long-forgotten legal interpretation.

By their nature, legal structures are meant to be universally applicable in a particular jurisdiction. This allows for strategic legal creativity: The claim against RWE stretched Section 1004 across a planetary scale. With surprising foresight, nineteenth-century German lawmakers recognized that we are all connected by a "sea of air." Pollutant imponderabilia can produce impacts across borders. While lawyers had previously applied Section 1004 to cases of local interference and pollution, the lawsuit drew on its drafters' original motivations and applied them to global climate change. Building on Section 1004, the claim reconfigured climate change as an engagement between neighbors.

Neighborhood Tensions in the Andes

Luciano Lliuya's claim caused reverberations around the world; many saw him as a hero of climate justice. However, among some of Luciano Lliuya's neighbors in his Andean village, the claim raised suspicion. In 2014, he was a farmer and mountain guide who lived in the village with his wife and two children. He and his family tended to the fields and cared for their animals. Starting when the claim was first announced in March 2015, his name periodically appeared in the Peruvian and international press following important legal events in the case.

At the lawsuit's outset, the directors of Germanwatch made a legal commitment to cover all costs associated with Luciano Lliuya's lawsuit.[13] The NGO raised funds from other organizations and private donors to cover legal costs and travel expenses and organize a public relations strategy for the lawsuit. Luciano Lliuya incurred few direct expenses. Germanwatch ensured that the court and lawyers received their fees.

After Luciano Lliuya's first visit to Germany in 2015 to submit the lawsuit, he and his family began renovations on their house in the flood hazard zone. The old adobe hut that his parents had built years before would hardly withstand a deadly flood wave. Working with extended family, they built a two-story brick house in its place. Luciano Lliuya used money he had earned as a mountain guide, but his neighbors speculated that he must have gained a significant financial windfall through his lawsuit and visit to Germany. Few confronted him directly, but rumors abounded about his supposed newfound fortune. Some even said he was selling Palcacocha to the Germans.

Luciano Lliuya did not explain to many people what he was doing. He lacked the charisma that allows others to hold dramatic speeches and draw people onto their side. He did not like addressing tense village assemblies. His neighbors found out about the claim primarily through gossip. Some younger people read about it on Facebook. When I began conducting ethnographic research in the area where Luciano Lliuya lives in 2017, I found that many people I spoke to did not understand why Luciano Lliuya had made the claim. Some had the feeling that he was doing something wrong. When I first moved in with a local family in a village neighboring Luciano Lliuya's, we sat around a table in their adobe house eating potato soup on a cold evening. Chatting with one of my hosts, I mentioned that I knew Saúl Luciano Lliuya. "Saúl is a crook," he exclaimed. "The people say that Saúl is making money from Lake Palcacocha." I explained the reasoning behind the claim to my host, but I still wonder to what extent he saw me, too, with suspicion at that point. To many, it simply seemed outlandish that Luciano Lliuya would go to Germany over a lake in the Peruvian highlands.

The irony was not lost on Luciano Lliuya: While he sought to establish a neighborly relation with RWE in Germany, the case caused friction between him and his neighbors in Peru. If neighborliness arises out of claims about mutual responsibility, it should come as little surprise that neighborly relations are not always friendly. In any social setting, the scope of neighborliness relates to people's moral understandings of how they should engage with one another.

Anthropologists have documented the significance of reciprocity for social relations in the Andes (Allen 1988). Local regulations in Luciano Lliuya's village and neighboring communities require people to participate in communal labor such as road and canal maintenance. While this custom is slowly declining as more people move to the city, communal works remain significant in the region as a symbolic practice that strengthens communities (Osorio Bautista 2013). Reciprocity also extends to relations between people and the landscape, which many engage as sentient. Across the Andes, people perform tribute payments to mountains and other sentient ecosystems to plea for more productive harvests (de la Cadena 2015). At Palcacocha, which threatens to flood the city of Huaraz, I witnessed locals providing tribute to the nearby mountains to prevent an avalanche and subsequent disaster. Mountains are neighbors of a different sort, traditionally seen as powerful beings requiring moral respect. Anthropologists have argued that climate change questions cosmological understandings about the power of mountain beings (Bolin 2009). Many feel that the mountains are suffering due to glacial retreat. They worry about what will happen to the mountains when they lose their white covers.

While reciprocity is a common theme in Andean notions of neighborliness, community relations can be contentious in practice. In my own experience, village assemblies are frequently tense affairs as people argue over how to divide access to community canals and how much they should pay for shared infrastructure services. Open arguments are not unusual. When conflicts arise between neighbors, they often concern matters of disputed mutual responsibility; a family might be accused, for example, of taking too much water out of the local irrigation canal.

These tensions are exacerbated by climate change, which increases rural Andeans' vulnerability. My interlocutors frequently discussed their concerns about melting glaciers, changing rain patterns, and water scarcity. Only some people referred to this as climate change (*cambio climático*), a term they picked up from friends, the media, or tourists. Climate change has no equivalent in Quechua, and many struggle to make sense of the situation. Locals see glaciers as an important source of water and are concerned about water scarcity as glaciers disappear. Recent research on climate change perceptions in the region has found that people provide varying explanations, including local littering, mining contamination, and even religious immorality (Rasmussen 2015). Some villagers have picked up on NGO and governmental discourses about environmental pollution and argue that people's incorrect behavior in their mountain range caused glacial retreat. They frame climate change as a local issue concerning relations between humans and the sentient Andean environment (Paerregaard 2013; Jurt et al. 2015). I encountered similar discussions in my research, with opinions differing about what caused the environment to change so drastically.

It did not help Luciano Lliuya's case with his neighbors that the lawsuit concerned the short-term risk of flooding rather than long-term water scarcity. Many in the area questioned authorities' assertions about the flood hazard at Palcacocha, arguing that this was a pretext to steal public funds through infrastructure projects. I encountered a pervading mistrust of state authorities among villagers in the rural areas around Huaraz. Most people simply assume that government officials are corrupt and are skeptical of anything they say. The government does not engage villagers as a reciprocal actor, and it frequently disappoints their expectations of support in terms of infrastructure and economic opportunities.

While the lawsuit against RWE is concerned with establishing causality and responsibility retroactively, many villagers would rather look to the future and wonder how they might continue to make a life for themselves. The legal claim addressed international discussions about climate justice but offered little concrete support to people who worry about the viability of agriculture for the next generation. For those who understand climate change as an issue of local environmental engagement, meaning that glacial retreat is caused by humans' disrespectful behavior toward sentient mountains, Luciano Lliuya's actions appeared completely misguided. From this perspective, RWE is not a relevant moral actor. To many in the village, the company is too far away to be a neighbor in terms of an individualized social relation. Locally, people usually use the Quechua term for "neighbor" only for those living in their immediate vicinity.[14] Meanwhile, German jurisprudence does not recognize mountains as legal persons. Luciano Lliuya found it difficult to reconcile his argument about global climate justice with some of his immediate neighbors' understandings of what it means to be a good neighbor and of which actions are viable for addressing glacial retreat. As his behavior did not make sense in the ethical order of neighborly relations, some came to perceive him as morally dubious.

It comes as little surprise that Luciano Lliuya faced rumors that he was doing something wrong. Luciano Lliuya speculated that many were simply envious of his newfound fame and attention. Envy and competition, particularly between men, are not uncommon in an Andean context marked by machismo. Based on a study of rural life in the Bolivian Andes, Krista van Vleet (2003) finds that gossip is a tool with which people make sense of relationships and social events. Gossiping is a theorizing practice that allows gossipers to evaluate people's behavior in relation to shared understandings about how community members should act. In this context, argues van Vleet, envy arises when someone is seen to gain an advantage without keeping to the moral obligations of reciprocity and sociality. As such, the critique of Luciano Lliuya arose from the perception that he had violated the moral order of Andean neighborly relations. These tensions may be declining; when I visited Huaraz in 2022 and 2024, many were more familiar with the lawsuit, and there seemed to be greater support for Luciano Lliuya in his community. An analytical focus on neighborliness, in the Andes as in the context of global climate politics, uncovers the moral stakes of social relations: Who are the relevant social actors, and how should they treat each other?

Litigation as Legal Strategy

Climate litigation brings political claims about climate change into the legal sphere by drawing people, organizations, corporations, and governments into a web of morally charged relations. In concrete terms, the claim against RWE enacts a neighborly relation between plaintiff and defendant. Like other climate change activists, Luciano Lliuya and his collaborators sought to promote social, political, and economic change addressing global warming. Luciano Lliuya, Verheyen, and Germanwatch came into the lawsuit with distinct yet overlapping aims. Asked about their objectives in press interviews, Verheyen argued that her goal was to achieve legal protection for her client against the harmful impacts of climate change. In her formal role as a lawyer, her duty was to defend her client's legal rights, and her focus was on the specific relation between plaintiff and defendant. The claim demanded that RWE pay a relatively small sum of money toward reducing glacial lake flood hazard in the Peruvian Andes. The fact that RWE refused to settle, instead presumably incurring millions in legal costs, highlights the claim's broader appeal to universal standards of moral engagement. In media interviews and public statements, Germanwatch representatives pointed to the lawsuit's potential for setting a legal precedent for holding large emitters responsible for their contribution to climate change. A positive verdict for Luciano Lliuya could provide the basis for future claims against other companies, in both Germany and other jurisdictions that have similar nuisance and liability laws. While a large number of lawsuits may be unlikely given the cost and effort required to bring forward each case, Germanwatch cofounder Klaus Milke would welcome additional litigation to make polluting corporations contribute to addressing climate change impacts: "This test case against RWE is already crucial for an equitable political approach to global warming and is also influencing the climate negotiations at the United Nations."

An increased risk of litigation can pressure policy makers to address the causes and impacts of climate change. At the 2022 UN Climate Summit in Egypt, governments decided to establish a global fund to help vulnerable countries address "Loss and Damage," which refers to the irreversible impacts of climate change such as glacial retreat and sea level rise. This agreement was achieved after decades of wrangling between vulnerable countries demanding compensation and wealthy countries refusing to accept any formal liability for climate change. Luciano Lliuya and I attended the summit with a team from Germanwatch, and he campaigned for a breakthrough on Loss and Damage at public events. Delegates told us that the threat of further climate litigation helped to build momentum in the negotiations. One representative said they raised the example of Luciano Lliuya's lawsuit with other negotiators, arguing that there would be many more such claims if they didn't come to an agreement on Loss and Damage. Questions remain as of 2025 about who will pay into the fund and how money will be distributed, but the agreement was a significant political victory for vulnerable countries demanding climate justice. Meanwhile, major energy companies have made significant climate commitments—at least on paper—with RWE planning to phase out coal and become carbon neutral by 2040, though the viability of these targets has been questioned (ClientEarth n.d.). As energy companies continue to make immense profits through burning and selling fossil fuels, climate litigation against major emitters is unlikely to subside anytime soon.

What did Luciano Lliuya want to achieve with the claim? Witnessing potentially devastating environmental transformations in his Andean home, he hoped to make a small contribution to slowing global warming and glacial retreat. He understood this is a long and difficult road and that, paradoxically, the claim provided little concrete support to help him and his community make a better life in times of unpredictable environmental change. He felt a responsibility to take whatever action he could, despite facing difficult repercussions at home. "I don't care what the other people in the village say," he told me one night in his house as we reflected on the lawsuit. His eyes welled with tears—"I'm doing it for the mountains."

The Limits of Law

Climate litigation translates broader political concerns into specific legally actionable claims. To achieve this transformation, lawyers and activists deploy law in new and unusual ways. As in many other climate litigation cases, Luciano Lliuya's lawyers used existing legal mechanisms that predate contemporary concerns about global warming. While the strategy can attract significant public attention, it also requires that activists fit their claims into judicial frameworks that restrict who can take action against whom and what kinds of claims can be made.

Regulatory frameworks determine who has legal standing, that is, which actors can make and defend claims. Only clearly defined legal entities can act in judicial proceedings, such as people, corporations, government institutions, and other legal persons. Climate change often affects larger communities in similar ways, as in the city of Huaraz, where around fifty thousand people inhabit the area affected by flood risk from Palcacocha. For legal and logistical purposes, a joint claim involving all fifty thousand people would be immensely complex and expensive under German law. It was much easier for Luciano Lliuya to bring the lawsuit forward on his own. Other potential actors, such as the Andean mountains, are excluded entirely. In addition, Luciano Lliuya could have sued numerous other polluting corporations alongside RWE, but doing so would have involved significant jurisdictional difficulties as those companies are based in many different countries.

A claim between one person and one corporation was easier to pursue within the judicial framework, yet it individualizes the problem of climate change. In terms of German neighborhood law, the case concerned the alleged violation of one person's property rights. Commenting on the case, Luciano Lliuya's lawyer argued that "private law involves balancing the rights of various entities in the real world, and Luciano Lliuya's rights should be enforced." This strategy potentially distracts from the systemic nature of global warming. As Greta Thunberg has eloquently argued, society and economic systems must change fundamentally to limit global warming and prevent significant socio-ecological disruption (Rowlatt 2020). At stake are relationships between human communities, environments, industry, and government institutions across spatial and temporal scales. While activists use litigation strategically to achieve broader aims, the formal scope of legal claims is necessarily limited to individual relations between specific actors.

The individual rights framing restricts what kinds of claims can be made and how they can be enforced. As discussed earlier, legal climate activists must fit their claims in a given judicial framework. Luciano Lliuya's lawsuit emerged after more than a decade of discussions among German jurists about how they might apply existing legal provisions in suing polluting corporations over their contribution to climate change impacts. The lawyer and legal scholar Wilhelm Frank first proposed in a 2010 article that German Civil Code Article 1004 might be used for this purpose (Frank 2010).[15] This provision permits only claims in which property ownership is clearly defined. Potential claim makers may face similar restrictions in other jurisdictions where formal property ownership is a prerequisite for legal standing. These judicial frameworks exclude claimants from places where land rights are disputed or regulated according to alternative standards. While in formal terms, the legal sphere brings together plaintiffs and defendants on equal footing, as anthropologists have argued, legal systems enshrine and reinforce asymmetrical power relations (Starr and Collier 1989). In the present day, many judicial systems disadvantage or exclude subaltern parties.

On his own, Luciano Lliuya had no access to the German judicial system. He was able to make the claim only with support from Germanwatch and his lawyers, as well as large financial contributions from donors to cover legal costs. The people at Germanwatch booked his flights, helped him obtain a visa, and housed him during his stay. Most people affected by climate change, especially in the Global South, are not in such a privileged position. They may have rights on paper but lack the connections and resources to enforce them. And while Luciano Lliuya trusted that the German courts would treat him fairly, he had less trust in the Peruvian judiciary. If legal systems are corruptible, rights have little value.

Judicial systems also limit what type of justice claimants can achieve. A legal victory for Luciano Lliuya would have provided him and his community with little immediate benefit, as any money he received would have gone to regional authorities as a minor contribution to a glacial lake hazard project. Legal success would be symbolic in the short term, while it would potentially contribute to political and socioeconomic change in the long term. Although the judge may have asked rhetorically whether it was just that people like Luciano Lliuya suffer the worst impacts of climate change while companies like RWE pollute the planet with impunity, the judicial framework offers only limited answers pertaining to specific neighborly relationships between individual actors. Even if people affected by climate change can have their rights recognized through legal claims, courts may be limited in their ability to enforce those rights. After a Dutch court ruled in 2021 that Shell must reduce its greenhouse gas emissions by 45 percent by 2030, the company moved its headquarters from the Netherlands to the United Kingdom—and the decision was overturned by a Dutch appeals court in 2024.

When climate change is brought to court, it is usually framed as a matter of individual rights. Legal claims invoke neighborliness in terms of specific moral relations between plaintiffs and defendants. Yet Luciano Lliuya and his interlocutors at Germanwatch aimed to set a precedent that would benefit many others. This aspect of the campaign was crucial for generating public backing and donations. Many cared about the claim because they hoped it could help achieve justice for more people. While the case was ultimately dismissed, it contributed to a growing legal foundation for holding corporate emitters accountable. Enforcing that foundation involves a new set of challenges. Each new lawsuit must again individualize climate change as a relation between a specific polluter and a specific impacted person. Activists hope that if companies face an immense wave of lawsuits, governments will finally take decisive action. Individualizing climate change as a legally constituted neighborly relation is a strategic move to draw attention and push for change at various levels of society. Yet, in the long term, limiting the devastating consequences of climate change for humanity and our planet requires solutions that benefit all members of the global neighborhood.

ON JUNE 30, 2017, the Upper State Court in Hamm made a historic preliminary ruling in favor of Luciano Lliuya. Reflecting their oral comments in the hearing earlier that month, the judges found that the lawsuit may proceed—it had a solid legal foundation and should not be dismissed. They saw the possibility for a precedent under German law if evidence could be found to prove a causal link between RWE's emissions and glacial lake flood risk to Luciano Lliuya's house in Peru. The decision was widely covered in German and international media, prompting discussions about whether major emitters should take financial responsibility for climate change impacts. That day, RWE's stock price declined by almost 2 percent, or by over €100 million.

Although the court ultimately dismissed the claim in 2025 on factual grounds, finding the flood risk insufficient to impose liability, it affirmed a landmark legal principle: Major corporate emitters can, in principle, be held responsible for climate-related harms. The ruling set an important precedent and will likely catalyze legal and political debates around climate accountability. Climate litigation continues to gain momentum worldwide. Time will tell how these efforts evolve in the courts and how they shape public conversations about responsibility for a warming planet.

Interlude 1

Andean Life in an Uncertain Climate

"Life isn't what it used to be," said Arturo Cacha.[1] "We worry for our future."

At the age of sixty-one, he looked like an old man. As he squinted in the midday sunshine, the wrinkles glistened on his face. A baseball cap covered the gray flurry of hair on his head. Standing in rubber boots and a well-worn shirt on a field rising steeply above his house, he grabbed his pickax to continue plowing the soil. Normally Cacha used his bulls to plow the fields, but this terrain was too steep. His entire body ached, but there was more work to do.

Cacha was tired. As usual, he had risen that morning at six a.m. He rose several times during the night to check on his cattle in the field adjacent to his house. Several neighbors had recently lost their livestock to thieves. His two bulls were still there, but we had heard the dogs barking at night—an indication of strange characters passing by.

Cacha lived in a small village above the city of Huaraz in Peru's Cordillera Blanca. His community neighbors Luciano Lliuya's. For around two months, I had lived in Cacha's house, working in the fields and eating with the family. "We're poor," Cacha sometimes said, and their circumstances are typical of many rural families in the region. They grow potatoes, corn, and other crops on their small plots of land. Their house is made of adobe bricks that provide insulation during cold nights but let in little light during the day. The women usually cook over open fires. The whole family works in the fields—plowing, planting, and harvesting year-round. Many young men have found paid employment elsewhere. Cacha's son works as a guide for mountain-climbing tourists during the dry season between April and September. Cacha worked with tourists in his younger years, but now he was too old and frail.

As he was climbing down a ladder that morning, Cacha's well-worn boots had lost their grip on the rungs. He crashed on his back and could hardly breathe, but nobody was nearby to help him. Limping over to the kitchen where his wife was preparing potato soup with fresh eggs, he felt a stabbing pain in his rib. Perhaps it was broken, but the doctor is expensive, and someone had to plow the field.

After breakfast that morning, Cacha and I set out with his three bulls and our pickaxes up a steep slope behind his house. He has spent his entire life working in the mountains. In recent years, the environment has been changing. "We used to have green pastures all over for our animals," Cacha explained while driving the cattle along with a little stick. "Now the animals get sick, so we have to buy medicine. You know, antibiotics. It's not like it used to be." As I gasped for breath under the morning sun, we reached the top of the hill and found a small patch of grass. Cacha set his pickax down to tie the animals' legs to a nearby tree. He spoke to me in Spanish, mixing in phrases in Quechua, a language I was beginning to learn.

Quechua was Cacha's first language. Growing up in the 1960s, he encountered a Peruvian educational system and society that gave little recognition to his native language and made him learn Spanish. To attend school, he had to walk several hours to reach the nearby city of Huaraz. Today, most villagers are bilingual in Quechua and Spanish, mainly speaking Quechua with their families and Spanish in Huaraz. Many young people prefer to speak Spanish among themselves. Quechua speakers continue to face discrimination from Peru's urban elites.

"It's all changing. It's warmer than it used to be." Cacha pointed to Mount Churup in the distance. Dark patches stood out under a shiny glacier that covered the mountain's peak. "Churup used to be completely white. Now the glacier is disappearing, and we don't know what will become of us in the future." Year after year, people in the Cordillera Blanca witness mountains losing their white covers. Many are concerned since they regard glaciers as a source of water. If the glaciers are gone, what will they drink?

"Life in the countryside is hard work," Cacha explained. We labored on beneath the burning sun, hacking at the dry soil with our pickaxes. Throughout the whole year, there is always something to do. Cacha worked in the fields since he was a child. Back then, his village was a *hacienda*, and the feudal system forced him to work for a wealthy landowner. In the course of the national land reform in 1969–70, his parents acquired several fields that they later divided among the family. With the younger men working in the mountains and the women attending to matters at home, Cacha took charge of preparing the fields.

Past midday, the heat became almost unbearable as we toiled on. During the Andean dry season, the nights are freezing and the days scorching hot under the sun. Sweating and struggling with pain, Cacha worked relentlessly. "Ease up," I advised him, as I feared going at it as he was would only make his injuries worse.

Cacha paused to observe our progress. We had plowed around a third of the little field. "Only a little more," he exclaimed.

Minutes later, we heard a whistle from the house below. Squinting in the sunlight, I made out Rocío Huamán,[2] Cacha's wife, waving to us. Cacha threw aside his pickax and turned to me. "Time to eat! Later I'll finish with the field."

We stumbled down the hill and sat down in the kitchen. It was refreshingly cool. After a morning of hard work, Huamán served us healthy portions of rice and beans. She had spent the morning cooking and washing clothes in a little tub. Her hands are rough, attesting to a life of hard labor in the house and in the fields. She too had felt an excruciating ache in her whole body for several months, but unlike her husband, she kept this fact quiet. Months later, when I had improved my Quechua and gained more confidence with her, she told me about her lifelong struggles as a wife and mother.

Like Cacha, Huamán witnessed the glaciers melting and worried about future water supply. For several years, their household enjoyed the benefits of tap water after government authorities laid pipe from a nearby mountain spring to supply their part of the village. But as the mountains lost their white caps, those springs began to dry up. The previous year, they had no tap water for several weeks during the dry season. Huamán and her daughter had to walk to the nearby river and carry buckets of water up the steep road for cooking and cleaning. With the rainy season now over, she feared water scarcity. "It's harder for the men to understand," she later explained to me; "it's us women who really know what it's like when we don't have water. We are the ones who suffer the most."

As we made our way back to the field after lunch, I asked about a large sack of potatoes standing in the kitchen. It went up to my chest and looked heavy. "Are those from the last harvest?"

"Yes," Cacha replied. "We used to store them for up to a year, but now the worms start eating them." Times have changed. The fields produce less than when he was younger, Cacha explained. With the warmer weather, new crop diseases have begun to appear. Cacha and his neighbors began using chemical fertilizers and pesticides to ensure better yields. Another farmer remarked to me that potatoes taste better without pesticides, but with the chemicals they could produce and sell more at the market.

Farmers are used to a regular cycle of seasons. They usually plant crops with the first rains in September. The rainy season used to start every November and last until around April. June to August were the driest months. The rains determine their agricultural cycle, allowing farmers to put food on the table. But in recent decades, this has changed. "The rains are no longer that reliable," Cacha told me. Sometimes they begin early; sometimes they come late. Farmers no longer know when to sow their fields. When the rains come, they often seem more intense than in previous years. Strong rainfall and hail can damage the crops. With increasingly unreliable rains, Cacha and his neighbors demand that government authorities improve irrigation infrastructure. Local politicians make ambitious promises during their election campaigns, but few deliver. Life is hard and the future uncertain.

Why were these changes happening? "The Earth is growing old," an old man in the village later told me. "We have entered a new time where the water, soil, and climate are changing." Villagers disagree about the causes of those changes. Some use moralistic arguments: Problems emerge in the environment because people treat one another and the world around them with disrespect. Some point to environmental contamination in the local biosphere, including plastic bottles that people throw into rivers and the tailings from nearby mining operations. Others, especially among the younger generation, speak in terms of global warming—big industry around the world is causing climate change. Reflecting these sentiments, villagers expressed an ambiguous mix of views about Luciano Lliuya's legal activism, ranging from support to suspicion. But what unites most farmers in the region is a concern about the mountains themselves, which are suffering as a result of glacial retreat.

3

The Politics of Personhood

The 2017 decision by the Upper State Court permitting the lawsuit *Luciano Lliuya v.* RWE to proceed led to a flurry of media reports around the world. In Germany, it prompted prominent commentaries on TV news programs calling for greenhouse gas emitters to be held accountable for climate change. Numerous journalists and documentary filmmakers visited Luciano Lliuya in Peru. Discussions emerged in Germany and beyond about who should take responsibility for climate change and how politics could respond to the challenge. The claim propelled Luciano Lliuya to international stardom, making him "a modern David" taking on the "Goliath" of RWE, according to a profile in *Time* magazine.

When Luciano Lliuya spoke about the mountains outside the courtroom, he brought an entirely foreign type of person to the legal process—one who had no formal standing or existence there. Nevertheless, his statement resonated with public audiences. Later that day, several German TV channels broadcast his words about justice having heard the mountains cry to millions of viewers on the evening news. His invocation of the Andean mountains raises an important question: Can the mountains play a role in the case, even if the court does not recognize them as participants? Formally, the lawsuit was between Luciano Lliuya and RWE. Both are considered to be persons under German law. Legal personhood gives them rights and responsibilities: Luciano Lliuya can claim that his rights have been infringed on, and RWE must respond in court. The Andean mountains were not officially involved, yet for Luciano Lliuya they are a powerful actor. In this chapter, I examine how corporations and mountains have emerged as different kinds of persons during Luciano Lliuya's legal action. Corporate personhood is a widespread legal concept crucial to the functioning of contemporary neoliberal capitalism. The personhood of mountains shapes how many rural Andeans engage with their environment. Both ideas of personhood involve ontological claims, meaning that they reflect a particular understanding about the nature of reality. In the lawsuit, RWE and the mountains became entangled in the politics of personhood. *Politics of personhood* refers to discussions about which actors have a stake in social, political, and environmental disputes. It revolves around the question of who counts as a relevant social actor (Walker-Crawford 2024).

This chapter explores how Luciano Lliuya and RWE came to count as potential neighbors in the legal process, while the mountains played no official role. Other research has shown how legal systems around the world exclude ecosystems as participants, though efforts are being made in some jurisdictions to grant them legal rights. I argue that ecosystems can play a role in judicial claims even without formal legal recognition. Tracing the construction of personhood in the claim between Luciano Lliuya and RWE, I show how both the company and Andean mountains were brought into existence as social actors when people engage with them. The claim hinged on the idea that RWE could be held liable as a single entity for emissions produced in numerous coal-fired power plants that operated across Europe over the past century. Luciano Lliuya and his supporters did not just accept the ontological notion of corporate personhood; they actively participated in making RWE a legal person by asserting the company's responsibility. This active stance allowed Luciano Lliuya to draw RWE into a legal relation. More broadly, Luciano Lliuya asserted that major greenhouse gas emitters are morally connected to those around the world who suffer the worst impacts of climate change. When Luciano Lliuya reminded everyone of the mountains outside the courtroom, he introduced the possibility that other persons might also be entangled in this web of moral relations. Corporations and mountains are made into social actors through ethical claims about climate justice.

Corporations, Mountains, and Other Persons

Personhood is a social phenomenon. Around the world, cultural and legal customs govern who counts as a person and what it means to be one.[1] Power relations shape the way people come to understand themselves as particular types of persons (Rose 1996). Historically, personhood has been a frequent subject of political and legal deliberation. In the US context, Christopher Stone (1972) traced disputes over whether foreigners, women, African Americans, Native Americans, and unborn fetuses should be considered legal persons. The notion that corporations are persons is a long-standing legal doctrine in many countries, allowing them to enter into contracts and face legal claims while protecting investors from personal financial liability (Blair 2013). In German jurisprudence, a corporate legal person is an organization that exists independently of its members and directors. It acquires rights and duties and can participate in judicial proceedings as a legal subject. As capitalist corporations gained increasing social significance during industrialization in the nineteenth century, they gradually came to be recognized as legal persons (Raiser 1999). Following a similar historical trajectory, US jurisprudence formally recognized corporations as legal persons in the late 1800s (Johnson 2012). Around the world, disputes are ongoing over what specific rights and responsibilities corporate personhood should involve.

Researchers have analyzed the social implications of corporate personhood, particularly since the controversial *Citizens United* ruling by the US Supreme Court in 2010, which granted free speech rights to corporations, allowing them to make unlimited financial contributions to political campaigns.[2] Corporate personhood is a wide-ranging concept that shapes how we think about both corporations and human persons. While numerous people and activities are often associated with a corporate entity, the notion of personhood gives corporations agency and accountability (Kirsch 2014). For Welker (2014), the idea of the corporation as a coherent actor reflects an individualist liberal model of subjectivity that defines humans as rational, self-interested actors. This definition makes it possible to identify the corporation as an intentional subject, though, in practice, the boundaries of the corporate person are often unclear. While some argue that corporate personhood is an illusion (Bashkow 2014), many lawyers and legal scholars engage the corporate person as "an organic social reality" (Blumberg 1990, 50). I argue that corporations become persons when people engage them as persons. In this way, the notion of corporate personhood is made and remade through mundane social practices, such as when corporations are brought to court as rights-holding legal defendants. This chapter examines that process ethnographically.

The notion that corporations are persons, able to act of their own accord, has become the subject of widespread critique, particularly from activists seeking to limit corporate power. Erin Fitz-Henry (2018) traces US environmentalists' efforts to promote legal rights for nature. Activists have sought to destabilize corporate personhood by bringing other potential persons into play, arguing that ecosystems should have more rights than corporations. For Fitz-Henry, this activism brings ontological questions about personhood into US legal and political discussions. *Ontology*, in this context, refers to conceptions about the nature of reality; for example, whether or not corporations and trees should be considered as rights-holding subjects. At the same time, new legal movements have emerged that support granting rights to ecosystems. In a seminal essay, Stone (1972) argued that if humans and corporations are considered legal persons with various specified rights, then natural environments should be protected under similar provisions. Decades later, such rights of nature have found official recognition in several jurisdictions. In 2008, Ecuador became the first country in the world to recognize rights of nature in its constitution (Gutmann 2021). In 2017, the New Zealand government recognized Mount Taranaki as a "legal personality," acknowledging Māori demands to apprehend the mountain as a being in its own right (Roy 2017). In 2019, the High Court of Bangladesh ruled that all rivers in the country are living beings with legal rights (Samuel 2019). Debates continue about how successful these legal changes are in terms of recognizing Indigenous perspectives and protecting ecosystems (Tănăsescu 2022).

Indigenous conceptions about sentient environments have also been the subject of recent anthropological discussions. Following the "ontological turn," authors have argued that researchers should take people's ontological claims seriously: Even if you do not see an Andean mountain as an actor in its own right, consider that it is a powerful entity for the people living below its peak. Most of this work focuses on the ontological standpoints of Indigenous peoples and other subjugated groups. For example, anthropologists have shown how people attribute personhood to animals and plants in Amazonia (Viveiros de Castro 2012) and palm trees in West Papua (Chao 2018), or how Andean mountains have appeared as potential participants in mining conflicts (Li 2015). Recent research in the Peruvian Andes has foregrounded the social significance of "earth beings," referring to a relational engagement with the environment that recognizes mountains, lakes, and other environmental features as living beings (de la Cadena 2015). When rural Andeans make tribute payments to landscapes, they draw them into a social relationship and enact them as sentient beings (Stensrud 2016b).

While courts have granted rights to ecosystems in some countries, most political and legal systems do not recognize them as legitimate actors. This leaves many people in the Andes and elsewhere feeling unrepresented. Sentient ecosystems are part of their daily lives but find no consideration in modern political institutions. Addressing this issue, de la Cadena (2015) has called for cosmopolitical dialogue: Modern politics should accept ontological difference. Bold (2019) calls for a cosmopolitics of climate change that acknowledges ecosystem actors. Climate change, in this sense, is an opportunity for cosmological conversation. I push this proposal a step further: Climate politics, as it played out in Luciano Lliuya's lawsuit, involves both sentient ecosystems and corporate persons that are brought into being through people's engagement with them. In the remainder of this chapter, I explore how Luciano Lliuya, RWE, and the Andean mountains emerged as potential participants in legal proceedings and in public discussions surrounding the claim.

Personhood Under German Law

When Luciano Lliuya first visited Germany in 2015 to file the lawsuit, we took him to see the RWE headquarters near the courthouse in Essen. As Luciano Lliuya glanced up at the imposing skyscraper of glass and steel, I said to him, "It's them you're suing."

"Really?" Luciano Lliuya grinned. "Don't scare me!"

It was his first physical confrontation with the company. Until that point, it had only existed to him as an idea; a potential target of litigation discussed on transatlantic video calls with German lawyers. Later that day, after filing the lawsuit at the courthouse, we visited a nearby open-pit coal mine operated by RWE. Luciano Lliuya stood in awe, gazing at the massive hole in the earth's surface that extended almost as far as the eye could see. He later explained to me that it felt like he was taking on something powerful.

Luciano Lliuya's suit arose in the German legal system, which sets strict limits on who can make claims, whom a claimant can approach, and what they can demand. Luciano Lliuya participated in the lawsuit because he felt the responsibility to act on behalf of Andean mountains facing devastating transformation. The claim asserted a neighborly relation between Luciano Lliuya and RWE as natural and legal persons. It rested on Section 1004 of the German Civil Code, a general nuisance provision that allows for claims arising from property interference.

How did Luciano Lliuya and RWE emerge as neighbors in the legal process? At the outset, Luciano Lliuya's lawyers had to provide documentary evidence proving the parties' existence. The claim said that "Mr. Luciano Lliuya is a natural person," and to it a copy of his national identity card was attached. An excerpt from the commercial register served the same purpose for RWE. To establish the company's liability, the claim stated: "Greenhouse gas emissions are primarily emitted by the defendant's subsidiary companies, particularly as a necessary consequence of coal-fired power generation. These emissions are to be attributed to the defendant juridical person as the parent company, particularly because the construction and operation of the power plants is not determined by the subsidiaries but occurs based on the defendant parent company's direction."

With this, the lawyers tied greenhouse gas emissions produced at numerous power plants across widespread locations to RWE. Configured as a legal person, RWE acquires an identity independent of its founders, owners, and employees. Lacking the individualistic confines of a human body, corporations are relational beings; RWE became enmeshed in neighborly relations that made it an agent subject to potential legal claims over its activities. Drawing on scientific research about climate change and flood risk in the Peruvian Andes, the lawyers sought to draw a causal link between Luciano Lliuya and RWE.

In ontological terms, the lawsuit enacted RWE as a person, engaging it as an independent entity subject to legal rights and obligations. In response to this engagement, lawyers representing RWE filed lengthy legal replies in the company's defense. The replies acknowledged RWE's personhood and asserted that it had the right to burn coal and should not be held legally liable for the impacts of climate change on Luciano Lliuya's life. In court hearings, judges addressed both Luciano Lliuya and RWE as judicial parties. The lawyers and judges enacted RWE as a legal person. Even to Luciano Lliuya, the company felt real and powerful when he walked by the towering headquarters on his way to file the lawsuit. Luciano Lliuya and his interlocutors engaged RWE strategically. From the perspective of Luciano Lliuya and his activist backers, the company served as a symbolic placeholder for the global polluting industries. By acknowledging the company as a defendant and commenting on its felt power, Luciano Lliuya participated in the social practices that make RWE a corporate person. The legal claim enacted both Luciano

Lliuya and RWE as particular types of persons, bringing them together as neighbors.

The lawsuit was successful in drawing an ethically charged relation between Luciano Lliuya and RWE. It brought him and the company's representative face-to-face in German courtrooms where their lawyers discussed the moral implications of global warming for major polluters. But what can be made of Luciano Lliuya's reference to the sentient Andean environment? The judicial framework grants no standing to mountains, and they found no mention in the legal documents. According to the politics of personhood enshrined in German law, RWE is a legal person susceptible to judicial claims from other recognized persons, such as Luciano Lliuya, while landscapes are not relevant actors. Nevertheless, Luciano Lliuya met an enthusiastic and emotive response when he spoke of the crying mountains in the courthouse lobby. His words appealed to widespread conceptions of Andean indigeneity. He introduced a new kind of person to the politics of climate change.

Powerful Mountains

On a cold, dark morning in July 2017, I accompanied Luciano Lliuya in climbing Mount Vallunaraju, overlooking the city of Huaraz. After sleeping in a tent at the foot of the glacier, we rose before dawn and set off. Attached to each other by rope, we held ice picks with thick-gloved hands and wore crampons on our boots to walk on the ice.

It was pitch-dark as we set off from the base camp. We wore headlamps to see. "This entire area—the campsite—was covered by the glacier just a few years ago," Luciano Lliuya told me. Every year that he climbed the mountains, he found the glaciers had receded by a few more meters. As we made our way onto the ice, the initial incline was relatively easy. It felt like walking up a snowy hill as the freezing wind hit my face. Then Luciano Lliuya, ahead of me, stopped walking. As I made my way up to stand beside him, he removed his gloves and pulled a little plastic bag out of his jacket pocket, leaving the ice pick dangling by a sling on his wrist. From the bag, he pulled out a handful of coca leaves and tossed them to the ground. "For the mountain," he said.

For many Andeans, engagement with powerful landscapes is a part of daily life. While ritualistic practices are often subtle in the Cordillera Blanca—as when Luciano Lliuya casually paid tribute to the mountain during our climb—they play a significant role for many people I met. Particularly in rural Quechua-speaking areas, people told me stories about powerful mountains and lakes that give life to farmers by providing water and fertility yet can respond violently when treated with disrespect. People's engagement with the environment seemed to be something very personal; it usually took time for me to build enough trust that they would share their experiences with me.

Luciano Lliuya is uncertain about what it means for the landscape to be alive. "A mountain is a geological formation," he later explained to me, "but another perspective is that the mountains nurture us. They used to be considered as gods—something to fear." Lying in his tent at night during climbing tours, Luciano Lliuya sometimes encounters the mountain he is about to surmount in his dreams, in the form of a person talking to him. He has also had experiences that are difficult to explain. On one tour, he woke up at night to the sound of voices and footsteps. Looking outside his tent, there was nobody to be seen. "I don't know; it's difficult to say with certainty," Luciano Lliuya explained, "but something is there." It may defy explanation, yet the significance is clear: "For me, the mountain is someone who gives you everything." A feeling of responsibility toward the Andean landscape was one of the primary reasons motivating Luciano Lliuya's participation in the lawsuit against RWE. Mountains are neighbors of a different sort. His engagement with them is shaped by a sense of moral duty; their power is unquestioned.

On another occasion during my research, I chatted with a group of villagers about how the environment was changing. An old man adjusted his wide-brimmed hat and surveyed the horizon. "What will happen to the mountain when the glacier disappears? I don't know—maybe it will die. A few years ago, the mountains were all white. Now look! The peaks are becoming dark." We sat by a small pasture in a village far above Huaraz in the Cordillera Blanca. I was visiting the village with Eduardo Díaz, who oversaw safety works at Palcacocha. Under the burning midmorning sun, the villagers expressed concern about potential water scarcity and increasingly strong hailstorms that threatened their harvests. I asked whether they engaged in any customs such as offerings to the mountains. Such practices were not common anymore, the old man replied; perhaps that was part of the problem. "The mountains are living beings, just like you and me; we have to respect them." Respect, for him, meant thanking the mountains for plentiful harvests and not leaving plastic bottles or other trash to litter the environment.

Stories about sentient mountains are passed down through the generations. Some speak of glacial lakes as enchanted places that can bewitch unassuming wanderers. The Quechua term for a snow-covered mountain is *raju*. The language lacks a clear distinction between a glacier and the mountain it rests upon, pointing to the value and significance of a mountain being as a whole.

Driving back down to Huaraz later that day, I turned to Díaz. With deep wrinkles on his face, he gazed out at the mountain landscape he called home. Working at Palcacocha, he made monthly offerings to the mountains and lake to keep them happy. Only that, he asserted, would prevent a devastating avalanche and flood. On cold nights in a little shack by the lake, they spoke to Díaz in his dreams. They demanded coca leaves, food, and drink. Díaz considered it his responsibility to maintain positive relations with the landscape.

"What do you think, Eduardo?" I asked as he turned to look at me. "What will become of the mountains when the glaciers melt off?"

Díaz paused to ponder the question, glancing past me toward the snowcapped peaks in the distance. "The mountains are hurting, that's clear. But I think they will be OK." He looked me in the eyes. "The Earth, the mountains, they're powerful. They always find a way."

Another villager from the area outlined his theory of how the mountains were changing—they had become calmer. "They used to be *chúkaro*," he explained, employing a Quechua term used to describe wild animals. "The mountains were more dangerous. But now that tourists come and climb the mountains every year, they've become tame." People may not fear the mountains as they used to, yet the landscape claims lives every year in mountain-climbing accidents. Some locals trace these deaths to a lack of respect for the mountains.

The environment is changing in uncertain ways, yet many villagers can feel its potency. "It's something quite intimate," Luciano Lliuya once explained to me. "When you're growing up, your parents tell you stories about powerful mountains and enchanted lakes, things like that. For example, they say that you shouldn't throw away food that you produce—it means the food will cry." According to Luciano Lliuya, the environment's sentience is self-evident for most villagers. "It's something you know—because you've lived all your life in those surroundings."

Through their relational engagements, Luciano Lliuya and his compatriots enact the mountains as powerful beings. In fact, Luciano Lliuya's own sense of self—of who he is and what responsibilities he has in relation to the world around him—emerge in relation to the Andean environment that comprises the people in his village and the sentient landscape. Much as the lawsuit involved a relational assertion that enacted Luciano Lliuya and RWE as human and legal persons, Luciano Lliuya and the mountains emerge as particular kinds of beings through his engagement with the landscape. While the relation between Luciano Lliuya and RWE involved legally inscribed categories of personhood, his and other Andeans' relations with the environment are more ambiguous. The people I spoke to have a strong sense of the mountains' power, yet they do not all agree about what this power means in practice. In addition, they are uncertain about how the mountains and their relationship with the landscape will change in times of global warming and glacial retreat.

Despite their ambiguity, Andean ecosystems are playing a part in legal and political disputes about environmental justice. Just as the legal invocation of RWE was a deliberate move to enable a politically charged claim, sentient landscapes can emerge strategically within the framework of activism. This strategic referencing occurred when Luciano Lliuya traveled to Germany to fight his case.

Strategic Indigeneity

On a sunny day in early November 2017, days before the yearly UN Climate Summit began in Bonn and around a week before the court hearing in Hamm, Luciano Lliuya stood on a stage in front of a crowd of twenty thousand cheering protesters (figure 3.1). It was a large turnout for a demonstration demanding immediate action on climate change from the world's governments. Locals were concerned about RWE's expanding coal mines in Germany, one of which had recently displaced a village and forest (Watts 2017). In an act of civil disobedience, some had even entered the open-pit mines and climbed RWE's giant diggers. Today, they joined with a crowd of foreign activists who had traveled to Germany for the UN summit.

"Since I was a small child, I could see changes in the mountains. Glaciers are retreating very quickly." Standing beside Luciano Lliuya at the podium, I translated his Spanish words into English. "We are very worried about the water, but this is something that we haven't caused. This was caused by the big companies that burn coal and petroleum. That's why I've sued the company RWE to hold them responsible."

A roar swept through the crowd. For the activists, Luciano Lliuya was a climate justice hero whom they could finally see in the flesh. He was a small man with a puffy red jacket, but his words carried significant weight.

FIGURE 3.1. Luciano Lliuya speaks to the climate protest in Bonn, November 2017. (Photo: Alexander Luna)

"We have the obligation, the responsibility to protect our Pachamama, which is the mother Earth, in the Andes." The audience cheered even louder. Luciano Lliuya had to wait for the crowd to quiet down before he could continue. His words resonated with many who saw him as an Indigenous activist who likely had a special tie to the landscape he inhabited—the Pachamama. Yet I translated this to the audience—using the widely recognizable "Pachamama" in English—with surprise. I had never heard Luciano Lliuya use the word.

Pachamama is a key term in multiple varieties of the Quechua language, including that spoken in the former Inca capital of Cusco. It translates roughly as "mother Earth," but often encapsulates a broader understanding of the Earth as a life-giving force, embedded in Andean cosmologies. However, the word is not widely used in the Cordillera Blanca. For Luciano Lliuya, *Pachamama* is a foreign term that he strategically applied to gain public resonance.

With this, Luciano Lliuya followed in the footsteps of other Andean environmental activists who have purposefully deployed ideas of Indigeneity. At a mining conflict in the Peruvian city of Cajamarca, north of Huaraz, protests achieved a new public dimension when activists began to describe the mountain under threat as sacred, using the word *apu*. This Quechua term, often used to describe agentive mountains, is common in other parts of Peru, but not in Cajamarca, where few people speak Quechua. While this terminology reflected locals' understanding of the mountain as an agentive being, it invoked popular imaginations of indigeneity that spoke to journalists and international activists. Some proponents of the mining project argued that this characterization of the mountain as *apu* was fraudulent, but Fabiana Li (2015) contends in an ethnography of the dispute that a binary opposition between "authentic" Indigenous tradition and invented conceptions is misguided. Even if locals had not commonly used the term *apu*, this conception opened political discussions to understandings of the landscape as agentive. According to one local activist, *apu* was a useful translation as the term had wider resonance, even if people in Cajamarca did not traditionally use it.

After the demonstration in Bonn, I asked Luciano Lliuya why he used the term *Pachamama*. "It's a nice concept," he replied. "I understand that Pachamama is mother Earth." Yet growing up in the rural Cordillera Blanca, he had heard a different set of stories. "My mother always told me, 'There's the sun, our father, and there's our mother, the moon.' But I don't remember her telling me anything about Pachamama." When he was thinking about his speech that morning over breakfast, he wanted to find something that would resonate with the audience: "Pachamama has already become widely known, and so they can understand it." Luciano Lliuya realized that many saw him as an Indigenous climate activist, even if he did not call himself Indigenous as the term is not widely used in his region. That positionality—of the Andean David facing the industrial Goliath—legitimized and strengthened his public image. At the demonstration, he drew on an Andean conception that was foreign to him yet served to support his public standing. "So you borrowed the term?" I asked.

"Exactly—because it was convenient and made sense."

When the German court took Luciano Lliuya's lawsuit seriously, it gave him a platform to bring other perspectives into conversation. Luciano Lliuya strategically applied a widely recognized "Indigenous" concept to raise this issue with the German and international public, even if that concept was not Indigenous to his own lifeworld. While doing so involved a simplification and repackaging of his complex and ambiguous engagement with the Andean environment, it served the strategic aim of showing a foreign audience that the environment itself might be concerned about global warming.

Luciano Lliuya has become a public figure, and his participation in a widely publicized legal trial lent public legitimacy to the idea that mountains and other nonhuman persons have a stake in legal and political discussions about climate change.

Mobilizing Personhood for Legal and Political Change

Luciano Lliuya engaged both RWE and the Andean landscape by asserting moral relationships with them. These connections involved ontological claims regarding who has a stake in contemporary political concerns regarding climate change. The company RWE and the Andean mountains emerge as distinct beings through claims about relational responsibility. With these engagements, Luciano Lliuya also emerged as a particular type of subject—both to public audiences and himself. Speaking outside the courtroom, he was an Indigenous activist seeking redress from industrial polluters. Offering coca to the mountain during a climbing tour, he was a rural Andean asserting a positive relation with the powerful landscape. Nevertheless, Luciano Lliuya's public assertions about sentient mountains contributed to growing claims that legal and political systems should account for ecosystems in their own right. The lawsuit invoked a neighborly relation between Luciano Lliuya and RWE as morally responsible persons. This proposes the idea that climate politics should account for such relations. It opens the possibility that other, nonhuman persons might also have a stake in these discussions.

Debates are ongoing about how ecosystem personhood and rights of nature can or should be recognized in existing legal systems. While a number of countries have recognized rights of nature, Andreas Fischer-Lescano (2020) recognizes persistent challenges in applying legal personhood to the ecosystem: Nonhuman legal persons cannot articulate for themselves, meaning they require representatives who must interpret their presumed interests. These interests may be difficult to define and may even conflict. Legal frameworks for representing corporate persons are well established: Corporations have governance structures such as boards and managements that express their interests. To account for ecosystem persons, legal frameworks may have to be rethought, argues Fischer-Lescano (2020). If they are badly applied, rights of nature may even harm Indigenous peoples whose perspectives they supposedly represent. Legal personhood rights were granted to the former Te Urewera national park in New Zealand in 2014, for example, ostensibly recognizing Indigenous claims to the territory. However, this conferral ultimately perpetuated conservationist policies that had historically dispossessed Indigenous peoples of their land. The environment's recognition as a legal person restricted Māori people's access to resources in the area (Coombes 2020). Whether or not ecosystem persons are adequately recognized by the law, they play an important role in the lives of many people facing climate change and environmental degradation in their communities. For Luciano Lliuya, the existence of mountain beings is self-evident, whether or not they are recognized under German law.

In the meantime, some people are strategically embracing notions of indigeneity to attract broader public support. Speaking to a crowd of protestors at the UN Climate Summit, Luciano Lliuya spoke of Pachamama. The concept is not native to his region, but it speaks to his ideas about the sentient environment. It was also a concept the crowd understood and could attach to him as a native Andean. Similarly, Indigenous groups in Ecuador and New Zealand have strategically embraced the idea of granting legal personhood to the ecosystem, even though the rights of nature concept did not fully represent their own philosophies and engagements with the environment (Tănăsescu 2020). Translating ideas across conceptual boundaries involves strategic compromises. When Andean ideas about the environment are brought to European courts or international political discussions, they must be communicated in a way that makes sense to a foreign audience.

Ecosystem persons can play a role in legal claims even when they have no formal standing. Luciano Lliuya brought the Andean mountains into play when he spoke about their relevance outside a court hearing in Germany. Over the years since, he has given numerous statements to the international media about how climate change is making the mountains suffer. Both the mountains and the defendant corporation are made real through people's engagement with them. The legal recognition of corporations as persons under modern capitalism is an "acknowledgment of the power of nonhuman actors that is made real and visible to humans through legal ritual performance," much as Indigenous ritual practices ascribe personhood to the ecosystem (Martin 2019). Ongoing discussions about rights of nature involve negotiations about what role Indigenous conceptions of the environment can play in contemporary legal systems (Tănăsescu 2020). These discussions do not just concern legal questions about who can participate in judicial processes; they ultimately revolve around political questions as to whose concerns should be taken into account and how humans should engage with the environment (Tănăsescu 2022). Legal action can draw new issues into the realm of the political (Eckert and Knöpfel 2020). When Luciano Lliuya spoke about the mountains in the context of his lawsuit, he raised the possibility that they could be considered as relevant actors, alongside humans and corporations, in legal and political discussions about climate change.

Part II

CAUSALITY IN THE COURTROOM

Climate change mitigation and measures to enhance climate protection are key elements of our corporate strategy.
—RWE, *Sustainability Report 2021*

It is undisputed that the potential climate risks of fossil fuels, due to the greenhouse gas emissions they release when combusted, have been widely recognized and reported on for decades.
—CHEVRON CORPORATION, *City and County of Honolulu v. Sunoco* LP

The importance of addressing climate change is not in dispute.
—ROYAL DUTCH SHELL, *Milieudefensie et al. v. Royal Dutch Shell* PLC

In a surprising turn of events, fossil fuel companies and climate justice activists now agree on the basic facts of global warming: Burning fossil fuels causes climate change, and humanity should make a concerted effort toward cleaner energy production. In recent litigation against major corporate emitters over their contribution to climate change, defendants have stepped away from overt climate change denial. Opinions diverge, however, on companies' responsibility to address the problem. While defendants admit that their actions played a role in global warming and claim to be part of the energy transition, they have rejected legal claims to reduce their emissions and accept financial liability for climate change impacts. Climate science is a key point of contention in legal arguments: Defendants accept its validity in understanding climate change and developing political responses; however, they question its feasibility in establishing legal responsibility. These discussions have high stakes: If courts recognize attribution science as legitimate evidence in establishing causal liability, major emitters could face a large number of lawsuits over climate change damages.

4

Truth and Responsibility in the Courtroom

"Now we come to the really interesting part," exclaimed Judge Meyer as he looked up from his notes and across the courtroom audience. During the hearing in November 2017 at the Upper State Court in Hamm, he had explained why Saúl Luciano Lliuya's claim against RWE was admissible in court. As a legal principle, the court found, it was possible for Luciano Lliuya to hold the company liable for its contribution to climate change impacts in Peru, and to demand that RWE contribute money toward infrastructural measures that would reduce the risk that Luciano Lliuya's house in the Andes would be flooded. This came as a great surprise to many in the courtroom. Looking over at RWE's lawyers from my seat at the plaintiff's table, I noticed that their eyes widened and eyebrows furrowed.

Judge Meyer continued: "RWE has argued that climate change requires political solutions. Political solutions are desirable, but irrelevant to the question of liability." The court needed to determine whether RWE caused an indirect nuisance to Luciano Lliuya's property through its greenhouse gas emissions via the process of global climate change. In legal terms, nuisance refers to an activity that interferes with someone's right to use their land. At the heart of the lawsuit was the issue of causality: Would the risk of flooding to Luciano Lliuya's house be lower without RWE's emissions? "Here," the judge explained, "that may be the case."

The RWE attorneys had entered the courtroom exuding the confidence and poise reflecting their client's and their firm's (Freshfields Bruckhaus Deringer) stature. Roda Verheyen, Luciano Lliuya's lawyer, had told him to keep his back straight and look self-assured—he had nothing to hide. Now, the five RWE lawyers began to sink into themselves, becoming ever smaller in their seats. Some of them wrote rapidly on their notepads and avoided looking up. Others kept touching their faces. On the plaintiff's side, Luciano Lliuya's lawyers exuded quiet delight. Luciano Lliuya later told me that he could feel the excitement in the room.

"Without going into the evidentiary phase," Judge Meyer continued, "we cannot exclude the possibility that RWE has contributed to the problem." The judges would require advice from court-appointed scientific experts to evaluate the alleged causal link between RWE's emissions and the risk of flooding to Luciano Lliuya's house. RWE's head lawyer held his head in his hands. When the judge later gave the floor to Verheyen, she was visibly overwhelmed: "You have left me speechless."

For many onlookers, the hearing felt monumental. Under German procedural law, courts move forward to hear evidence only if they are convinced that legal arguments have been conclusively resolved. Speaking to the press afterward, Verheyen described it as a historic moment. For the first time anywhere in the world, a court had declared admissible a claim to hold a polluting company liable for its contribution to climate change impacts. In legal terms, the judges found, the claim was possible. But to win, the lawyers would have to prove a causal chain linking RWE and Luciano Lliuya. The case was an attempt to draw a legal relationship between defendant and plaintiff as neighbors, a connection founded on scientific insights about climate change. To determine whether RWE could be held liable for its contribution to climate change impacts in Peru, judges in the Upper State Court in Hamm sought to answer the following evidentiary questions: Did the plaintiff's property face a "serious threat of impairment" due to a risk of glacial lake outburst flood from Palcacocha? Were there "scientifically verifiable facts" that pointed to a "serious threat of danger"? Was there a "sufficient probability that damage will occur within a foreseeable timeframe"? If this was found to be the case, could the threat be linked to RWE's emissions since 1958 via the process of global climate change?[1] The court ruled that the causal relation between fossil fuel combustion, CO_2 emissions, and global warming had been reasonably foreseeable since 1958, when the scientist Charles D. Keeling began publishing research on the anthropogenic greenhouse gas effect.[2]

RWE's lawyers vehemently denied the possibility of establishing a causal link between the company and Luciano Lliuya, questioning scientific evidence presented by the plaintiff. Yet, throughout the legal proceedings, a key set of facts remained undisputed: All those involved agreed that climate change is an anthropogenic problem that requires public attention. Undoubtedly, the company had produced greenhouse gas emissions through its operation of coal-fired power plants. Nevertheless, RWE's lawyers rejected the lawsuit on two fundamental levels: In normative terms, polluters should not be held liable for the indirect and unintended consequences of their emissions. In epistemological terms, they denied that it was possible to draw a causal link between RWE's emissions and an alleged risk of flooding affecting Luciano Lliuya's property.

In legal and political discussions, tracing causal responsibility for climate change is not merely a scientific question of measurement and modeling. It concerns the norms that should govern relations between people, institutions, and environments on a warming planet. Ethically charged neighborly relations are enacted through claims about causality. Causal evidence provides the factual basis for establishing a neighborly relation between plaintiff and defendant. Depending on its framing, climate science can be used to assert a link between polluters and impacted people at an individual or global scale, or to deny such a link altogether. The following chapters examine the normative ideas about climate change at stake in legal discussions about evidence. Scholars in sociolegal studies and science and technology studies (STS) have examined how science is used as evidence in the courtroom. I bring this work into conversation with recent scholarly debates exploring the role of climate science in claims against major polluters. In the judicial process involving Luciano Lliuya and RWE, legal practitioners deployed scientific facts to establish and contest causation. While most academic literature has focused on plaintiffs' perspectives, I delve into arguments presented by both sides. I show how different types of evidence come to count in legal proceedings. Meanwhile, litigants' ability to produce evidence is influenced by broader relations of power that shape the generation of knowledge about climate change.

Part 2 of this book traces the disputes about causation and evidence in *Luciano Lliuya v. RWE*. I begin with an overview of the ways lawyers use scientific evidence to make legal arguments. Luciano Lliuya's legal team sought to prove that the company caused a nuisance to Luciano Lliuya by contributing to a "serious threat of impairment" affecting his property. Luciano Lliuya's lawyers presented evidence that linked him to RWE through a complex causal chain: First, RWE produced greenhouse gas emissions through coal firing that contributed to global warming. Second, global warming led to glacial retreat in Peru. Third, glacial retreat in the Andes led to an increased risk of flooding affecting Luciano Lliuya's property. For each of these steps, lawyers argued over whether a link could be established and what kinds of evidence should count. An examination of the causal chain highlights how neighborly relations are enacted in legal practice, entangling human and corporate persons in ethically charged relations.

Legal Causation: Connecting Actors to Impacts

To determine whether a causal relation could be proven between RWE and Luciano Lliuya, the judges overseeing the case had to clarify the facts. Some facts appeared straightforward, such as the size of Luciano Lliuya's property and the year it entered his possession. Most facts in the legal process were disputed: Was there an immediate danger to Luciano Lliuya's property? Can RWE's contribution to glacial retreat in Peru be determined? In deciding what to accept as truth, the judges had to balance competing factual claims with legal principles of adjudication and morality. Their role tasked them with deciding which claims were most useful for resolving the question of causality.

A brief statement on the second page of Luciano Lliuya's lawsuit summarized the facts from his perspective: "[The claimant's] property is acutely threatened by glacial retreat, which is occurring as a direct consequence of climate change with increasing speed and magnitude."[3] This factual claim, placed prominently in an initial summary of the legal reasoning, included several key assertions:

1. Luciano Lliuya is a living person;
2. He legally owns a residential property;
3. This property is situated geographically in a place that leaves it exposed to an acute threat of flooding;
4. This acute threat of flooding is a consequence of glacial retreat;
5. Glacial retreat is increasing in speed and magnitude;
6. Glacial retreat is a consequence of global anthropogenic climate change.

The following paragraph in the lawsuit argued that RWE has contributed to anthropogenic climate change with its emissions. For each of these factual claims, the lawsuit included further substantiation and evidence. In the legal process, these six knowledge claims were subject to varying degrees of contention. Claims 1 and 2 were among the few that RWE's lawyers did not counter, though even those required proof. All claims were specific to the lawsuit's judicial framing: They supported the position that RWE was partially responsible for Luciano Lliuya's predicament and should contribute financially to adaptation measures. They emerged in relation to broader considerations among concerned citizens, activists, and lawyers about the social meaning and ethical implications of climate change.

The legal proceedings revolved around the question of causation: To what extent had RWE caused a risk to Luciano Lliuya's property? In environmental law cases, the causal attribution of human conduct in a particular outcome is often a core issue. To establish liability, lawyers must prove that a specific legal actor—whether a human or a corporation—acted in a way that contributed causally to a specific harm (or risk of harm) affecting another legal actor. The lawsuit between Luciano Lliuya and RWE was a civil dispute over liability for potential harm. In German and many civil law traditions, the causation test often draws on the doctrine of *conditio sine qua non*, which closely parallels the common law's "but for" test: *X* caused *Y* if, *but for X*, *Y* would not have occurred (Young et al. 2004, 509).[4] For example, person A drops their heavy bag on person B's foot, causing one of B's toes to break. Person A is causally responsible: But for A dropping the bag, person B's toe would not be broken. In climate change cases, causation is more complex, involving a causal chain that extends across the planet's surface and atmosphere. Causation, in that case, is cumulative; it involves multiple causal parties whose actions collectively result in global climate change, which, in turn, manifests in a multitude of specific local impacts.

In a journal article published shortly before the lawsuit was filed, Luciano Lliuya's lawyer Roda Verheyen (2015) established the conceptual groundwork. In claims over loss and damage associated with climate change, plaintiffs must overcome three hurdles to establish legal causation: First, they need to choose a specific defendant among the multitude of global polluters and determine their individual contribution. Second, plaintiffs must link a specific environmental change or event—such as glacial retreat or sea level rise—to anthropogenic climate change. Third, they must link a specific loss to that change or event. Figure 4.1 visualizes the causal chain in *Luciano*

FIGURE 4.1. The causal chain depicting the neighborly connection between RWE and Luciano Lliuya.

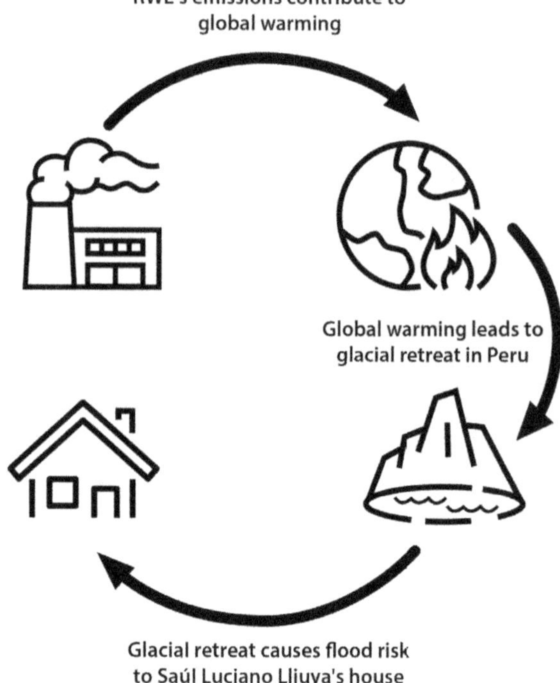

Lliuya v. RWE. The aim for Luciano Lliuya's lawyers was to link RWE to Luciano Lliuya by proving that the company contributed to a "serious threat of impairment" affecting the plaintiff's property. The first step was to link RWE's greenhouse gas emissions from its coal-fired power plants (top left) to global atmospheric warming (top right) by determining RWE's precise contribution. The second step was to connect global atmospheric warming to glacial retreat in the Andes (bottom right), demonstrated through climate change attribution science. Third, lawyers sought to link glacial retreat with a risk of flooding affecting Luciano Lliuya's house (bottom left), using flood modeling at a local scale. The following chapters walk through each link of the causal chain.

The judicial notion of causation, underpinned with scientific evidence, allowed Luciano Lliuya's lawyers to construct a legal and social relationship between a human person and corporate entity that transcended space and time: The two legal parties were located on different continents, and the lawsuit concerned RWE's greenhouse gas emissions since before Luciano Lliuya was born. In the legal documents, the academic and abstract arguments relating to causation drew on scientific data about global climatic processes. Nevertheless, the legal claim was successful in creating a direct relationship between the two legal parties, reflected in the courtroom confrontation between Luciano Lliuya's and RWE's lawyers. Legal causation arguments linking RWE to flood risk in Peru gave Luciano Lliuya and his supporters an opportunity to make a broader statement: that relations between polluters and those who face the worst impacts of climate change should be a key issue in social and political discussions.

Science in the Courtroom

During the legal proceedings in the Upper State Court in Hamm, the judges had to determine whether RWE contributed causally to a "serious threat of impairment" affecting Luciano Lliuya's property. If the judges in Hamm found that there was no serious risk to Luciano Lliuya's house, the case would be over. When legal parties submit evidence in court, they do not merely present indisputable truths. Rather, they draw on various types of knowledge to strategically construct factual claims that support their normative view about the issues under dispute. Courts evaluate these factual claims in relation to legal standards of proof, and when they decide to accept particular facts, these become inscribed as legal truth in rulings and verdicts. A key issue of concern for sociolegal scholars is how evidence is produced and brought to bear on legal arguments (Stehr and Weiler 2008; Schauer 2022).

Scientific knowledge often plays a significant role as evidence in legal trials. The discipline of STS provides helpful perspectives for analyzing how science and law interact. The value of these perspectives has been recognized by other scholars in the sociolegal literature who draw on theories and concepts from STS (Cole and Bertenthal 2017). For example, researchers have used STS approaches to study how legal knowledge is produced and validated in the context of litigation over urban crime (Valverde 2005) and criminal trials of environmental activists (Hayes 2013). The work of Bruno Latour and Sheila Jasanoff is particularly insightful in this regard. Latour (2010) has argued that judicial process, like science, is a regime of truth production: Its institutional aim is to discern justified claims from false assertions. Judges require facts to make legitimate rulings. In this framework, individual facts are tools that contribute to the legal system's overarching goal of producing justice. In a similar vein, Jasanoff (2005) has argued that for both science and law, the perceived absence of truth threatens their institutional legitimacy. Law differs from science in that it requires clear definitions of true and false (Latour 2010). Scientific inquiry is typically open-ended—scientific facts are potentially subject to future revision. Legal proceedings, on the other hand, usually need to establish the truth in the normative context of legal adjudication to allow for a verdict (Jasanoff 2005; Patton and Barnes 2017).[5] Judges seek to set the facts straight so that they can close the file (Latour 2010, 211). Judges and juries determine what becomes legal fact and which arguments fail to meet the threshold.

Scientific knowledge enters the courtroom not as bare facts or truth but as evidence (Haack 2014). Lawyers must present scientific knowledge in a format that fits the epistemological standards of law (Jasanoff 2006). Courts rely on scientific evidence to establish causation, interpreting it through the lens of legal reasoning (Stuart-Smith et al. 2021a). Since the Industrial Revolution, legal scholars and practitioners have held the view that scientific knowledge is one of the most reliable types of evidence for legal disputes (Jasanoff 2006, 330). Nevertheless, science and law often involve different standards of truth. While science strives to produce universally valid knowledge, law produces knowledge that is relevant within the confines of a particular legal case and jurisdiction. Often, legal processes involve different conceptions of facticity and truth than those required for scientific knowledge. What counts as a fact in a court of law may not count as a fact for science, and vice versa (Jasanoff 2007, 762). Law depends on nonlegal knowledges that enter legal cases as evidence (von Schnitzler 2014). For example, scientific evidence can set the parameters for the issue the court seeks to resolve, showing the court what is relevant and where it can intervene. While legal provisions set standards for truth finding in the courtroom, this process is often informed by scientific experts who are called as witnesses. Every jurisdiction has its own legal standards for deciding what counts as true, and these can differ between types of cases. For example, the standard of proof is typically higher in criminal law than in civil law.

When judges decide which facts are true, they actively participate in the process of legal knowledge production. Science does not provide facts in a legal sense; judges must measure scientific insights with the legal yardstick. Setting the framework for determining truth, "the law actively constructs the scientific facts that it presumes to 'find'" (Jasanoff 2007, 776). The legal process produces new knowledge. Whether in science or law, factual claims often relate to social and ethical conceptions of how socio-environmental relations are—or should be—organized. All facts are partial to the context of their production. Knowledge claims emerge hand in hand with conceptions about how the world is and toward what kinds of social relations people should strive. Knowledge and society are symbiotic, constantly reproducing and reinforcing each other in a process that Jasanoff (2004) terms "coproduction." The broader issues at stake became clear during Luciano Lliuya's court hearing in November 2017, when Judge Meyer suggested that the plaintiff and defendant might seek a settlement out of court. After all, the lawsuit was over US$18,800, a fraction of the mounting costs for legal proceedings and scientific studies. "No," replied RWE's head lawyer. "This is a matter of precedent." For both parties, this suit was about much more than a causal chain between Luciano Lliuya and RWE; it concerned relations between a much wider set of polluters and impacted people.

Numerous scholars in the sociolegal literature have studied how legal knowledge claims are linked to broader social concerns, particularly in the context of legal mobilization. In their analysis of legal mobilization against US military bases in Japan and South Korea, Claudia Kim and Celeste Arrington (2023) argue that activists use the judicial process as a site of knowledge production. Plaintiffs commissioned scientific research to measure noise pollution around bases and presented the results to courts. When judges accepted those studies as evidence, they gave "a degree of official recognition" to the findings. In a study of legal mobilization against racialized policing practices in France, Magda Boutros (2022) shows how activists used statistical studies pointing to racial disparities in police stops to argue that there was an issue of systemic discrimination. In a Botswana legal case brought by Indigenous groups over their right to live in a Kalahari Desert game reserve, discussions revolved around whether the plaintiffs could be considered "Bushmen," which status would entitle them to constitutional protections. The plaintiffs had been forcibly relocated from the reserve and prohibited from engaging in traditional hunting practices. In the trial, lawyers drew on anthropological literature and oral histories to argue that the plaintiffs should be considered Bushmen in both cultural and legal terms, giving them a right to continue residing in the reserve. In response, government lawyers drew on different anthropological research to argue that the plaintiffs' ethnic groups had assimilated into Botswana society and could not be considered as true Bushmen, denying them legal protections (Sapignoli 2017). Across these cases, analysis traces how plaintiffs strategically produce and present evidence that contributes to broader activist narratives.

Scholars studying climate litigation from a sociolegal perspective have employed Jasanoff's concept of coproduction to examine the link between legal knowledge and broader social concerns. Surveys have shown that many scientists' motivation to study climate change is linked to their personal concern about the issue (Milman et al. 2017; Tollefson 2021). This concern, in turn, is also a response to policy makers' demands of science to improve the knowledge framework for developing political responses to climate change (Hulme 2010b). When climate science is subjected to legal evidentiary standards, knowledge emerges hand in hand with social and political claims about how climate change should be addressed. Reviewing academic discussions about climate litigation from a sociolegal perspective, Elizabeth Fisher (2013) has found that litigation can legitimize broader social concerns about climate change. When climate change becomes a legal fact, it comes to appear more "real" in the public eye. Many cases have drawn significant public and media attention, turning courtrooms into sites of coproduction where activists draw on scientific research to make social and political arguments about climate justice. Graeme Hayes (2013) examined a prominent case tried in a British court in 2008 in which six Greenpeace activists accused of causing criminal interference at the coal-fired Kingsnorth power station in southern England successfully justified their actions with reference to global warming. The court heard testimony from scientific experts who drew a link between global climate change and local greenhouse gas emissions linked to the power plant, leading to the activists' acquittal. Hayes argues that expert knowledge claims in this case were coproduced alongside broader narratives of climate change and social justice, providing social and political justification for illegal actions. In the courtroom setting, knowledge and sociopolitical claims about climate change emerged hand in hand.

In a comparative study of claims against the Dutch, Norwegian, and Irish governments over climate policy, Phillip Paiement (2020) examines how the cases are embedded in wider transnational concerns. He finds that the claimants employ a common narrative about the clarity of science and urgency of government action. Analyzing these lawsuits in their broader social context highlights how factual claims in court are coproduced together with political claims about who should take responsibility for climate change. Much like Paiement, Lisa Vanhala (2020) traces how environmental activists have drawn on scientific research to advocate for the legal designation of the polar bear as "endangered," a strategy that opened new legal possibilities for advocating climate action to protect polar bears' habitats. Studies like Paiement's and Vanhala's point to the need to explore knowledge in its social context and reflect the recognition in anthropological research that knowledge is inherently relational and imbued with value (Ingold 2000; Strathern 2018). Ultimately, courts must decide whose facts are more factual.

Judicial standards of evidence specify how courts should decide which facts are most truthful. In Germany, judges enjoy a high degree of autonomy in legal decision making. Section 286 of the German Code of Civil Procedure sets the parameters of evidence for courts, regulating how judges should determine whether evidentiary facts are true or untrue: "*The court is to decide, at its discretion and conviction*, and taking account of the entire content of the hearings and the results obtained by evidence being taken, if any, *whether an allegation as to fact is to be deemed true or untrue*. The judgment is to set out the reasons informing the conviction of the judges" (Bundesamt für Justiz 2013a [emphasis added]).

Accordingly, German judges have significant leeway in deciding on the facts of a case. While other legislation sets standards for evidentiary permissibility in particular circumstances, judges can weigh different types of evidence. German jurisprudence has recognized that there may not always be absolute certainty about the facts: "In cases of uncertainty, the judge can and must make do with a degree of certainty that is serviceable to practical life; that silences doubters without excluding them entirely."[6] Rather than striving to arrive at absolute truth, the epistemological framework of German law sets the guidelines for judges to establish truths that are good enough for ruling on the issue at hand. Arguing within this framework, lawyers for Luciano Lliuya and RWE have sought to convince the judges with factual claims that either affirm or deny a causal link between the company and climate change impacts in Peru.

Law and Science on a Warming Planet

Legal frameworks determine which types of evidence are considered valid. The following chapters trace how evidence was produced and deployed in the trial between Luciano Lliuya and RWE. As in numerous other jurisdictions, evidentiary standards in the German judicial system grant significant value to scientific knowledge and formal expertise. Faced with contradictory evidentiary claims as to whether RWE contributed to a risk of flooding affecting Luciano Lliuya's property in Peru, judges in the Hamm court appointed independent scientific experts to provide their opinions and guide judicial decision making. What are the wider consequences of the centrality of science in climate litigation?

Scientific research can be expensive. A significant hurdle for campaigners interested in climate litigation is the cost of compiling scientific evidence.

Large emitters such as RWE have significantly more resources at their disposal to produce counterevidence, placing people affected by climate change, litigating activists, and environmental organizations at a disadvantage. In addition, the availability of scientific research limits the feasibility of new claims. For Luciano Lliuya's lawyers, a case involving the Peruvian Cordillera Blanca was advantageous as numerous studies about climate change impacts had already been conducted in the area, providing a solid evidentiary base. On a global scale, Christian Huggel and colleagues (2016) have pointed to a distributive injustice in climate change research as impacts in the Global North have been studied significantly more than those in the Global South. This uneven availability of scientific research creates additional disadvantages for potential claimants from the Global South. Lawyers and judges may also struggle to understand the intricacies of climate research and to translate scientific insights into a legal framework. In many climate lawsuits against major polluters, evidence submitted has lagged behind the scientific state of the art (Stuart-Smith et al. 2021a).

In legal proceedings, furthermore, framing a socially contentious issue such as climate change in terms of a scientific dispute may distract from the underlying normative concerns at stake. Judges defer to scientific expertise under the assumption that it can provide neutral and disinterested answers to resolve legal disagreements. This deference may obscure the fact that science is itself a highly normative endeavor (Latour 1987). In the present age, some researchers studying the processes and impacts of climate change may even be motivated by social and political concerns (Kotcher et al. 2017), seeking to produce knowledge that will contribute to political—and even legal—strategies for tackling global warming. Scientific disputes that arise at the heart of climate litigation involve not mere technical disagreement but normative concerns about how people should engage with one another and who should take responsibility for ongoing ecological transformations.

A final reflection on the centrality of science in climate litigation relates to public understandings of evidence beyond the courtroom. Litigation can provide legitimacy to concerns about climate change (Fisher 2013). Attribution science, which models links between localized events or processes and anthropogenic climate change, played a key role in the trial between Luciano Lliuya and RWE, along with research that quantifies individual companies' contributions to climate change. In these comparatively new fields of study, scientists are demonstrating at increasingly small scales that observable events such as the retreat of individual glaciers are related to global warming. Luciano Lliuya's lawsuit was one of the first cases in which attribution science was used in court to hold a major emitter responsible for an individual impact. If judges recognize such research as valid in legal proceedings against polluting corporations, they will create an immense legal precedent. They would also provide public legitimacy to the field of attribution science. This, in turn, could shape political discussions: If publicly recognized evidence points to the responsibility of industry for specific climate change impacts, politicians may face increasing calls to hold corporations financially accountable. As such, legal process has the potential to shape public and political standards of evidence.

Interlude 2

Courtroom Interrogation

"You're the person who traveled the farthest to arrive here," said Judge Rolf Meyer as he looked Saúl Luciano Lliuya in the eyes. It was November 2017, three years after Luciano Lliuya had decided to take legal action and two years after filing the lawsuit. He sat next to his lawyers in a courtroom full of mostly sympathetic spectators. The hearing had just begun.

As the judge addressed him, Luciano Lliuya's face froze in a moment of wide-eyed astonishment. His eyebrows shot upward, and his jaw dropped slightly. A court-appointed interpreter sitting next to Luciano Lliuya quickly relayed to him the judge's words. The day before, Luciano Lliuya's lawyer had told him it was unlikely that the judges would ask him to speak in court. In her long experience of environmental court trials, the lawyers did most of the talking. This judge took a more informal approach.

"Did you have a good trip? How are things in Germany?"

Luciano Lliuya sat up, maintaining eye contact with Judge Meyer, but his expression started to relax. "Good, thank you, the Germans are very nice." The interpreter repeated his words in German.

The judge began to ask a series of factual questions.

"What size is your property?"

"About one hundred square meters," Luciano Lliuya replied.

Luciano Lliuya and the interpreter had difficulty understanding each other. At times, she incorrectly interpreted the judge's questions in Spanish and incorrectly interpreted Luciano Lliuya's answers in German. Despite the difficulties, the judge carried on his conversation with Luciano Lliuya.

"When and how did you acquire the property?"

"It was my family's property. My father bought it."

Luciano Lliuya still appeared nervous, giving short and concise answers. As the discussion continued, he warmed up to the situation.

"But when did your father buy it?"

"Uh," Luciano Lliuya stuttered, "1979, or maybe 1980, around then." Luciano Lliuya's father bought the property in 1984, but Luciano Lliuya misremembered the date under pressure.

"And since 2014 you have owned the house?"

"Yes. My parents gave it to the children. Now I own it with my wife."

I was surprised that Judge Meyer asked about such simple factual matters. All this information was in the legal briefs that the judges had likely read in great detail. Was he perhaps trying to assess Luciano Lliuya's character? To test his authenticity?

5

Tracing Emissions

Two and a half hours into the court hearing in November 2017, Judge Meyer turned to the practicalities of evidence. He had clarified the court's opinion that the lawsuit was legally admissible. The judges found no reason to exclude RWE from legal liability in principle. If the claimant could prove a causal link between RWE's emissions and flood risk to his house in the Andes, the judges would rule in Luciano Lliuya's favor. RWE could become the first company to be held legally accountable for its contribution to climate change. Judge Meyer had rebuffed repeated and increasingly acrimonious attacks on the plaintiff's claim from RWE's lawyers.

"If we go into the evidentiary phase," Judge Meyer explained, "we will require independent experts who have not lost their impartiality." In the German legal system, entering the evidentiary phase meant that the court would appoint expert witnesses to evaluate the facts.[1] If need be, the experts could conduct additional studies to determine whether RWE was partially responsible for flood risk in Peru. "We will need a geologist and a climate scientist," Judge Meyer elaborated. He directed a cold look toward RWE's lawyers, who were shaking their heads and grimacing. "I don't understand what all the fuss is about."

I suspect that Judge Meyer understood well what the fuss was about. The defendant's lawyers were clearly unhappy that the court found the lawsuit to be admissible. Nevertheless, the judge appeared determined to press on to questions of evidence. After clarifying the next steps—the parties would present written arguments about the evidence—Judge Meyer ended the session. It had lasted for almost three hours. Later that month, the court issued a formal ruling that the case could enter the evidentiary phase. The court appointed two scientific researchers to examine the issue of flood risk in August 2018.[2] If the court found, based on the experts' testimony, that the flood risk reached the legal threshold, the judges would appoint a second panel of experts to evaluate whether RWE could be causally linked to the flood risk in Peru.

All those involved concurred on a crucial set of facts: Climate change is a problem of broad concern. Greenhouse gas emissions, including those from RWE's coal-fired power plants, have contributed to climate change. Yet the devil lay in the detail. While the fact of climate change remained uncontroversial, the lawyers argued over the precise nature of climate change. In this chapter, I unpack legal arguments concerning the first step of the causal chain: the link between RWE's greenhouse gas emissions and global warming. Building on anthropological discussions about the conceptualization of pollution within the legal framework, I show how the two parties offered different analytical frameworks for understanding the social-material-atmospheric processes of climate change. Lawyers for both sides deployed authoritative scientific and legal knowledge that made RWE's contribution to global warming appear either quantifiable or impossible to grasp. In sociolegal claims about climate justice, the very nature of climate change is at stake—in the ways it connects people, corporations, and CO_2 molecules ascending into the atmosphere.

Following the Particles

In a gripping legal ethnography of a lawsuit against Chevron over environmental contamination linked to oil production in the Ecuadorian Amazon, Suzana Sawyer (2022) describes how the company's lawyers argued in an Ecuadorian court that there was no health risk for local communities despite the unquestioned fact that the environment was full of crude oil residue. The lawyers applied an industry-sanctioned understanding of scientific risk analysis: They argued that even though soil and water samples showed high levels of petroleum hydrocarbon particles, they did not contain certain particles known to be toxic. Chevron's lawyers applied a perspective that rendered petroleum residue harmless. Following this industry-endorsed logic, plaintiffs would have to prove the presence of specific toxic particles to establish liability. In Sawyer's view, corporate risk management science established such strict specificity requirements for pollution that it was difficult to establish causation.

Chevron ultimately lost the case in an Ecuadoran court, though the company left the country before the verdict could be enforced. Sawyer's analysis of the Ecuadoran oil spill points to a key issue in legal environmental disputes: how courts qualify causality and harm depends on which knowledge frameworks they use to understand the problem. Climate litigation concerns complex relationships between greenhouse gas emissions, global warming, and local impacts. Legal parties present different approaches for understanding these relationships, giving rise to conflicting notions of responsibility and legal liability. All participants in Luciano Lliuya's lawsuit undoubtedly accepted that coal firing at power plants owned by RWE led to the emission of CO_2 and other greenhouse gas particles. Nobody doubted that many of these particles subsequently entered the atmosphere and contributed to global warming. Nevertheless, each side offered a distinct understanding of when and how greenhouse gas particles should be considered legally relevant.

The lawyers for RWE argued that it was scientifically and legally impossible to trace emissions to an individual polluter in cases of cumulative causality.[3] They drew an analogy to a parallel set of cases from the 1980s concerning damage to forests due to sulfur dioxide (SO_2) emissions, allegedly emanating from nearby industry.[4] The German Federal Court of Justice ruled that liability could not be established as numerous actors had emitted SO_2 that subsequently mixed in the air, making it impossible to determine whose molecules had damaged which specific trees. Citing this decision, RWE's lawyers argued that it was not possible to establish causal liability in cases of cumulative environmental damage. Much like SO_2 molecules from different sources mix in the air, CO_2 and other greenhouse gas molecules become inextricably comingled when they enter the atmosphere. Consequently, the RWE lawyers argued, there could be no individualized causal relation in legal terms between RWE and Luciano Lliuya or, more specifically, between RWE's emissions and potential climate change impacts in Peru affecting Luciano Lliuya.[5] The defendant thus focused on the materiality of individual greenhouse gas molecules as they ascend into the atmosphere and become lost among countless other molecules from countless other sources.

In response, Luciano Lliuya's lawyers highlighted the differences in behavior between SO_2 and CO_2 molecules. They pointed out that SO_2 molecules remain closer to the earth's surface, potentially causing damage by coming into direct contact with material environments. In terms of legal causation, this distinction placed a burden on forest owners to prove where the SO_2 molecules that damaged their trees had originated. Carbon dioxide molecules, on the other hand, collectively rise into the atmosphere, contributing to global warming. And, although they become inextricably mixed, every molecule reinforces the planetary warming process. Accordingly, the lawyers concluded, individual emissions can be causally linked to climate change impacts; unlike SO_2, CO_2 does not have to be linked by specific molecules with specific damages, as all emissions contribute to global warming.[6]

Luciano Lliuya's lawyer summed up this argument in an article: "Every molecule that is emitted, irrespective of where it actually comes from, contributes at least marginally to the greenhouse effect and thereby to the rise in temperature and its consequences" (Verheyen 2015, 164). This approach, focusing on the cumulative processes of greenhouse gas emissions and climate change, allowed lawyers to construct a neighborly relationship between Luciano Lliuya and RWE. To make these molecular processes causally legible for a legal liability claim, the lawyers established a framework for measuring the defendant's emissions in relation to all other emissions occurring on the planet.

Quantifying Corporate Emissions

In the early 2000s, the burgeoning group of climate change lawyers faced a challenge: there was a lack of evidence for private climate liability claims. The proposition linking specific companies and their emissions to specific climate change impacts lacked a strong scientific basis. Working with a group of environmental lawyers, Roda Verheyen and her colleagues commissioned a study that analyzed historical fossil fuel extraction and emissions. These efforts resulted in the Carbon Majors Report (Heede 2014a, 2014b). Led by the US geographer Richard Heede in cooperation with other academics, this research quantified industrial greenhouse gas emissions since industrialization and linked them to specific entities. It concluded that ninety companies are responsible for around two-thirds of historic industrial emissions.[7] The study was a key piece of contentious evidence throughout the legal process.

The company RWE was founded in 1898 at the height of German industrialization. As of its 2023 annual report, the company still relied on coal-fired power (RWE 2024). Consequently, it has become one of the largest greenhouse gas emitters in Europe. According to the Carbon Majors Report, RWE is responsible for 0.47 percent of industrial anthropogenic emissions between 1854 and 2010 (Heede 2014a).[8] Although RWE in its 2016 annual report called itself "the single largest carbon dioxide (CO_2) emitter in Europe" (RWE 2017),[9] worldwide it trails emitters such as ExxonMobil (3.21%) and Saudi Aramco (3.17%). The company has no presence in Peru. It has produced emissions primarily in Germany and in other European countries. In the United Kingdom, RWE owns a subsidiary called npower. The Carbon Majors study marked a significant turning point for climate litigation in that it made specific greenhouse gas contributions identifiable and measurable. Furthermore, it linked them to potential defendants. It provided a scientific basis for Luciano Lliuya's lawyers to argue that RWE should be held liable for 0.47 percent of the costs to reduce the risk of flood to Luciano Lliuya's house, in line with the company's alleged contribution to global industrial emissions.[10]

In their response to the lawsuit, RWE's lawyers questioned the Carbon Majors Report's scientific validity.[11] They argued that it was "not tenable," as its source material was unclear and incomplete. Furthermore, the report included a significant uncertainty factor, considered only industrial carbon dioxide and methane emissions, and insufficiently accounted for the changing ownership of specific power plants. These points relate to the discussion of corporate personhood in chapter 3. As some coal-fired power plants had been previously operated by other companies before being acquired by RWE, the lawyers argued, not all emissions from those plants could be linked to RWE as a legally responsible person. They highlighted two phenomena not considered in the Carbon Majors study: First, some carbon dioxide particles do not rise into the atmosphere and are captured by carbon sinks, such as forests, while some methane molecules break down before rising into the atmosphere. Second, they criticized the study's focus on industrial emissions, ignoring greenhouse gases produced through agriculture. Overall, the lawyers stated, climate change involves "a highly complex interplay between numerous factors and interactions that are characterized by high uncertainty and are not adequately understood to this day."[12] Therefore, they concluded, it was not possible to establish a causal link between RWE's emissions and climate change impacts in Peru.

Making Climate Change Legally Relevant

At the outset, both the lawsuit and RWE's initial response to it set forth why climate change should—or should not be—a relevant topic of discussion in a German court. The lawsuit painted a picture of climate change as a global process with distinct identifiable causes and impacts. It drew on German law, scientific publications, and Peruvian government declarations as authoritative sources of knowledge to argue that RWE should be held legally liable for its contribution to the flood risk affecting Luciano Lliuya's property in Peru. The suit stated that glacial melt and the growth of Palcacocha—which create the flood hazard—were "caused at least inter alia by anthropogenic climate change."[13]

The suit drew on one of the most authoritative sources of knowledge in a German courtroom: German law. Citing a law that regulates emissions trading for greenhouse gases,[14] the claim stated that the "existence of global climate change through increased concentrations of greenhouse gases such as carbon dioxide in the atmosphere is undisputed in Germany." Next, the document cited a public statement from RWE that acknowledged the existence of anthropogenic climate change in a discussion of the company's long-term goal of climate-neutral electricity production. It was "common knowledge," the lawyers conclude, that climate change is caused by increased atmospheric greenhouse gas concentrations.[15]

RWE's legal response did not question that climate change exists or that RWE's emissions contributed to global warming. Citing UN reports and numerous articles from prestigious scientific journals, RWE's lawyers painted an alternative picture of global climate change. From the company's perspective, the processes of exchange between people, the global environment, and atmosphere are extremely complex and not fully understood by the scientific community. It concluded that a causal scientific and legal link could not be established between RWE's emissions and a potential flood hazard to Luciano Lliuya's property.[16] The lawyers argued that climate change and its potential impacts could not be addressed in terms of corporate legal liability but required governmental solutions at a national and international level.[17] In terms of nuisance law, they found, climate change was not a relevant issue.

Climate change can acquire competing social meanings (Callison 2014). Lawyers for Luciano Lliuya and RWE contested the meaning of climate change through the judicial process. The climate emerges as a statistical abstraction and social fact through socially contingent relationships between people, technologies, knowledges, and the environment (Demeritt 2001, 312). Scientific knowledge is key, as it allows people to conceptualize the dynamics of climate change and make claims about how we should resolve the problems it causes. According to Callison (2014, 23), scientific discussions about climate change bring up complex questions with no simple scientific answers: How should we live with our planet and with one another? How do we relate local circumstances to global processes? The legal dispute between Luciano Lliuya and RWE was an attempt to address these questions via the normative framework of German law. Lawyers for the plaintiff and defendant offered distinct perspectives on causation and responsibility in the context of climate change. Each approach offered epistemological credence to the lawyers' attempts to make or unmake a material and moral relation between Luciano Lliuya and RWE.

Having reviewed the arguments about RWE's contribution to global warming, I now move to the next step in the attribution chain: the link between global climate change and glacial retreat in the Peruvian Andes. More specifically, Luciano Lliuya's legal team had to prove that anthropogenic climate change caused a specific glacier situated above Luciano Lliuya's hometown to retreat. Legal discussions on this issue involved disputes over the validity of climate models as well as over attribution science, which links individual impacts to climate change. Climate change knowledge emerges in the context of an international institutional framework that promotes and disseminates policy-relevant research about global warming. Scientific climate change models are now becoming enveloped in legal claims over climate justice.

Interlude 3

Climate Skeptics at Large

In September 2022, a new piece of climate research appeared on the internet: an article entitled "Attribution of Modern Andean Glacier Mass Loss Requires Successful Hindcast of Pre-Industrial Glacier Changes" in the *Journal of South American Earth Sciences* (Lüning et al. 2022). Two of the authors are Sebastian Lüning and Fritz Vahrenholt, who are well-known German climate skeptics. In 2013 they coauthored the book *The Neglected Sun: Why the Sun Precludes Climate Catastrophe*, which claims that global warming is due to the sun, rather than human activity. The 2022 article, which is four pages long, questions widely accepted methods and research in climate change attribution science pointing to the fact that anthropogenic climate change has caused significant glacial retreat in the Peruvian Andes. The article was published via open access which costs authors US$2940 as of 2025, plus tax.

In a response to the article, several experts in the field of climate change attribution stated that a table in the article includes manipulated data and that some of the references are inaccurate (Stuart-Smith et al. 2023). Two of the authors used to work for RWE. Fritz Vahrenholt was CEO of the RWE subsidiary RWE Innogy from 2008 to 2012, and then he was a member of RWE's board until 2014. Sebastian Lüning worked for another RWE subsidiary, the oil and gas company RWE Dea, from 2007 to 2012. The 2022 journal article was initially published online as an unformatted preproof version. Under the "Declaration of Competing Interest" section, the authors wrote the following: "Note that SL [Sebastian Lüning] and FV [Fritz Vahrenholt] are former employees of the company RWE; the study has not been influenced by this."

Why did two former RWE employees publish a piece of climate skeptic research that clearly favors RWE's position in the ongoing trial with Saúl Luciano Lliuya? In response to a media inquiry, RWE's spokesperson stated: "We did neither commission that study nor play any role in producing it" (Parry 2023). At present, the company has no climate deniers on its board or running any subsidiaries. Neither did their lawyers submit the specious article as evidence in the court in Hamm.

6

Modeling the Global Climate

Early climate litigation cases against corporate polluters in the 2000s faced difficulties as plaintiffs were unable to prove a link between emissions and impacts in terms of legal causation. Since then, climate change science has evolved rapidly. New studies and research frameworks make it easier to link specific emitters with specific impacts, providing a scientific basis for new claims and increasing the likelihood that future lawsuits will succeed (Stuart-Smith et al. 2022). Science is key to understanding climate change: It makes the global climate knowable (Sayre 2012). Vital in this regard is the detection and attribution of climate change impacts, a field commonly known as attribution science, which involves attributing individual impacts to global anthropogenic warming. Attribution science plays a significant role in climate litigation. Scientific evidence is critical for legitimizing claims in a legal context and establishing a causal link (Setzer and Vanhala 2019). Litigants draw on this new scientific framework to characterize climate change in terms of specific causal relations between emitters and impacts, rather than a global process with diffuse local impacts. Attribution science provides key evidence for constructing specific neighborly relations between major polluters and those who face the impacts of climate change.

When scientists build climate models, they make assumptions about how certain human and environmental factors have developed (Wynne 2010). Some researchers have critiqued the contemporary scientific and policy approach which assumes that anthropogenic impacts on the climate can be identified. According to Nathan Sayre (2012), the term *anthropogenic* arose out of a modern scientific conception that draws a conceptual boundary between humanity and natural environments and assumes that the latter are inherently stable without human influence. This disregards the fact that natural and human processes are inherently entangled (Hulme 2010a). If we cannot separate human actions from environmental processes, we preclude legal and social claims that tie specific polluters to particular climate change effects. Nevertheless, the study of human interference in the climate has significant ethical contours. The increasing impacts of human activity in all parts of the world are making it more and more difficult to distinguish between nature and humanity. Sayre (2012) rejects the potential universalizing implications embodied in the term *anthropogenic*—that humanity as a whole is transforming the planet. Rather, he calls for research about which people have caused which changes and who is affected. Recent advances in attribution science offer a step in this direction to the extent that they help establish causal liability in climate litigation against major emitters.

This chapter follows scientific discussions in court regarding the second part of the evidentiary chain: the link between anthropogenic global warming and glacial retreat in Peru. Lawyers argued about whether attribution science and climate models are good enough to hold up as legal evidence. I place RWE's legal defense in the context of broader efforts by fossil fuel companies to question climate science and delay climate action. The legal process was so slow that scientific research advanced and new attribution evidence emerged while the case was ongoing. Computer-generated models allow scientists to represent the complex atmospheric processes of climate change based on statistical and quantitative data. Since the 1980s, researchers have developed models that render the climate as an integrated system on a planetary scale, contributing to an understanding of Earth as a global environment (Miller 2004). In response to public and policy concerns about climate change, academics have developed climate models and attribution research to understand the problem and provide a scientific basis for developing solutions. Now, this research is becoming socially and politically charged in new ways as it provides the basis for legal accountability claims.

Climate Modeling on Trial

Lawyers for both Luciano Lliuya and RWE extensively cited reports from the Intergovernmental Panel on Climate Change (IPCC). The UN created the IPCC in 1988 to study global climate change and provide a scientific basis for the emerging international policy framework. As an intergovernmental body, the IPCC counts the world's governments as its members, along with leading scientists across academic disciplines. In its Assessment Reports, the IPCC summarizes and synthesizes the current state of academic research on the causes and impacts of climate change and the actions that would mitigate its impacts.

In the initial complaint submitted in 2015, Luciano Lliuya's lawyers devoted much attention to the parameters of scientific and legal truth, arguing that statements from the IPCC about climate change impacts should provide sufficient evidence for proving a causal link of accountability. Another section in the lawsuit explained the legal requirements for evidence. Citing German procedural law and past rulings defining the degree of certainty needed to establish that a fact is true, Luciano Lliuya's lawyer contended that claims defined by the IPCC as having a "very high probability" should be considered as true within the legal framework. While the IPCC acknowledges that all climate models and statistics retain a degree of uncertainty, the lawyers argued that this uncertainty—in cases where it remains relatively low—does not make a model's overall results unusable for German jurisprudence.[1]

To draw a causal link between global anthropogenic climate change and glacial retreat in Peru, the lawsuit cited the IPCC's *Fifth Assessment Report* (*AR5*), which establishes as "very likely" the fact that over half of the global temperature increase between 1951 and 2010 can be attributed to anthropogenic greenhouse gases (IPCC 2013, 932). Another section of the report states, with "high confidence based on high agreement and robust evidence," that glaciers have retreated at a rapid pace, particularly since the 1970s, in several countries including Peru (IPCC 2014, 1518–20). Citing the same chapter of the IPCC report (IPCC 2014, 1544), the lawsuit argued that glacial retreat in the Cordillera Blanca could be attributed to anthropogenic influence.[2]

In response, RWE's lawyers cited the IPCC's *AR5* to counter the lawsuit's claim that greenhouse gases have caused global climate change via an increase in global temperatures. Arguing that this conception was reductive, they stated that "global" climate change should not be considered synonymous with "anthropogenic" climate change. Rather, numerous natural and anthropogenic factors shape the climate. The response followed this premise with a long discussion of climatic drivers, including anthropogenic and natural greenhouse gases, solar radiation, aerosols and volcanos, land use and agriculture, as well as ocean cycles.[3]

Furthermore, RWE's lawyers denied that an increased atmospheric concentration of greenhouse gases led to a recent accelerated glacial retreat in the Peruvian Cordillera Blanca region: "Such a simplistic causal link, as the plaintiff wants to assume, does not exist." Relying on data from *AR5*, the lawyers argued that the global increase in greenhouse gas emissions was not linked in a linear fashion to global average temperatures: Emissions increases did not always correspond directly to temperature increases. Comparing data from *AR5* with a scientific publication about temperature development in the Cordillera Blanca, they argued that global average temperatures do not necessarily correspond with local temperatures measured in the Cordillera Blanca. While one increases, the other may even decrease. They concluded that "greenhouse gases emitted since the 1980s cannot have contributed to glacial melting."[4]

New Evidence

The lawsuit was dismissed on legal grounds by the Essen State Court in 2016. The dismissal was subsequently appealed to the Upper State Court in Hamm, where judges found the case to be legally admissible. Judges appointed scientific expert witnesses to help them address the evidentiary questions and planned a site visit to Palcacocha in Peru to gather evidence. While the legal process moved forward slowly amid procedural delays, and while the COVID-19 pandemic postponed the site visit, climate science evolved at a surprisingly quick pace. In 2021 an independent study was published that attributed flood risk at Palcacocha to anthropogenic climate change. Relying on climate and glacier modeling, it examined the relation among global atmospheric warming, local warming at Palcacocha, the retreat of the Palcaraju glacier, and a potential increase in flood risk. It found that around 95 percent of the local warming and subsequent glacial retreat was due to human influence (Stuart-Smith et al. 2021b). The study was published by researchers at the Universities of Oxford and Washington in the leading journal *Nature Geoscience*. It relied on established methods in the attribution community but was the first of its kind in that it examined the influence of climate change on glacial retreat and flood risk at a single location.

The study's publication caused a media stir. *The Guardian* ran a report and the lead author, Rupert Stuart-Smith, gave a long interview on BBC radio about how his research could be used as evidence in the *Luciano Lliuya* trial. One article called the study a "smoking gun" (Berwyn 2021). Luciano Lliuya's legal team was delighted: They had considered commissioning an attribution study for Palcacocha in the early stages of the lawsuit but decided against it. It would have cost around $100,000, and the lawyers assumed that the court would commission its own research if the case got that far. Now, a group of scientists from elite institutions had conducted an attribution study on their own accord. The results were not surprising, as earlier research had linked glacial retreat in the Andes to anthropogenic climate change, but it provided much more detailed evidence.

The article was submitted as evidence by the plaintiff's legal team in April 2021.[5] An accompanying legal brief, which I helped draft, summarized the study's conclusions for the judges. It explained that the study found a "clear causal link" between anthropogenic climate change, glacial retreat at Palcacocha, and a serious risk of flooding. It employed "scientifically recognized and widely used methods." The brief further elaborated that the article passed *Nature Geoscience*'s rigorous peer review process, meaning that it met scientific state-of-the-art standards.

RWE's lawyers took eight months, an unusually long time, to respond. They found the study to be unsuitable for resolving the evidentiary questions at stake. In a legal brief filed in December 2021, they stated that the study's methods were not adequate for establishing glacial lake flood risk, that it relied on insufficient data to prove a local temperature increase, and that its glacial modeling was imprecise. They also attacked the authors' credibility, an argument I examine below.

In a first line of critique, the lawyers argued that the publication offered "no new insights regarding a concrete GLOF [glacial lake outburst flood] risk at the Palcacocha glacial lake."[6] According to the defense analysis, the underlying methods were suitable only for identifying potentially dangerous glacial lakes at a larger scale and did not sufficiently account for an individual lake's parameters. For example, possible trigger events such as avalanches and landslides were not evaluated. As the article's primary focus was on attributing glacial retreat to global warming via climate modeling, it referenced more detailed studies that had modeled potential outburst flood scenarios.[7] RWE had questioned the scientific integrity of those studies in a previous legal brief, arguing that their flood models involved unsuitable and arbitrary assumptions.[8] The lawyers supported their arguments related to flood risk with specially commissioned research: a report by scientists at the university RWTH Aachen (Rheinisch-Westfälische Technische Hochschule), which receives funding from RWE. The report was authored by Florian Amann and Holger Schüttrumpf, who also served as expert witnesses for RWE in the trial. It carried RWTH Aachen branding and was marked "Not intended for publication" on the title page.[9]

As they did in the suit's early stages, RWE's lawyers questioned the efficacy of climate models for determining a local temperature increase in the Peruvian Andes. They argued that the model used by Stuart-Smith and colleagues (2021b) operated at much too large a scale to allow for reliable conclusions about temperature variability at Palcacocha. To prove an increase in temperature, they contended, the plaintiff must present measurement data from the lake. Where such data are unavailable, as in the case at hand, climate scientists typically use atmospheric models to simulate temperature changes. According to the lawyers, such modeling does not provide suitable evidence. They argued that the authors' approach had significant underlying uncertainties due to insufficient real-world data input and relied on imprecise assumptions. While not explicitly rejecting established climate science methodologies as a whole, the lawyers raised doubts about individual studies and datasets in order to undermine the plaintiff's evidentiary basis. They emphasized uncertainty to claim that existing research does not meet legal standards of evidence.

Similarly, the defendant's lawyers questioned the efficacy of a glacier model used by Stuart-Smith and colleagues (2021b) to evaluate how the Palcaraju glacier above Palcacocha has retreated since 1880. Once again, they pointed to underlying uncertainties and assumptions. The lawyers did not deny that the glacier has retreated—this has been documented in extensive research and historical photography—but argued that discrepancies between modeling results and actual measurements prevented the model in question from accurately representing how the glacier has changed. They concluded that "even if such models may be a relevant instrument for science, they do not fulfil the legal standard for causation and evidence." The model provided only "a statement of probability that is insufficient for establishing causal proof under civil law."[10]

Scientific Credibility on Trial

After the study by Stuart-Smith and colleagues (2021b) was submitted as evidence, the judges found in a ruling that it was an independent piece of research that had higher evidentiary validity than scientific studies commissioned by one of the parties to the suit.[11] While the article's conclusions were pertinent to the lawsuit against RWE, the study was conducted independently of the plaintiff's legal team, which were not involved in the process. A news report published by the University of Oxford (2021) stated that the study could help clarify the facts in the ongoing legal process. In a legal brief, RWE's lawyers questioned the independence and impartiality of the study's authors.[12] They alleged links between the study's authors and the plaintiff's legal team. They claimed that the authors sought to influence the lawsuit in the plaintiff's favor. In procedural terms, they argued, it was not independent proof; its evidentiary legitimacy was compromised.[13]

In addition, the lawyers alleged that some of the article's authors had "publicly positioned themselves as enemies of CO_2 emitters and have been active on the side of climate activists and climate litigators." They based this accusation on the fact that Rupert Stuart-Smith, the study's lead author, had expressed his wish to bring "together science and the law to tackle climate change." In addition, they pointed out, he regularly shared articles and opinions on his Twitter account about the responsibility of corporations in relation to climate litigation. Without providing evidence, they argued that Stuart-Smith was a member of the "United Kingdom Youth Climate Council," which, they alleged, organized activities with the activist movement Extinction Rebellion. They also asserted that another of the authors, Myles Allen, provided support to plaintiffs in a US climate lawsuit against the fossil fuel industry and authored an article in *The Guardian* titled "Big Oil Must Pay for Climate Change." For the lawyers, this was sufficient evidence that the authors' scientific work was biased against fossil fuel companies such as RWE.

The lawyers further questioned the authors' scientific integrity by associating them with the plaintiff's lawyers and scientific advisers. "The authors," they stated, "are part of a certain circle of people who present at events together, publish together and cite each other in publications." They pointed out that the plaintiff's lawyer Roda Verheyen and the author Myles Allen both contributed to an inquiry into climate change by the Philippines Human Rights Commission. Without providing evidence, they alleged that Stuart-Smith and his colleagues had contact with two scientists who advised the plaintiff's legal team.

This was not the first time RWE questioned the credibility of climate scientists. After the lawsuit entered the evidentiary stage in 2017, the court asked both parties to suggest scientists who could act as independent court-appointed experts. The court's preference was to appoint German-speaking experts. The plaintiff's lawyers proposed Friederike Otto, a well-known climate scientist who helped develop climate change attribution methodologies widely used in the scientific community.

RWE's lawyers countered that Otto was unsuitable as an independent expert because of an alleged bias against fossil fuel companies underlying her work. Their legal brief included color prints of three Twitter posts going years back in which Otto shared news articles about climate litigation. While Otto stated in her posts that the claims were interesting, the lawyers took this to mean that she "expressed support for legal claims in relation to climate change." In the same brief, the lawyers pointed out that Otto had participated in two events alongside lawyers from ClientEarth, an organization involved in climate litigation including claims against fossil fuel companies. Consequently, they found that Otto was trying to "position her work in connection to potential climate liability claims" by promoting it to environmental lawyers.

Finally, RWE's lawyers pointed to the fact that Otto and the plaintiff's lead lawyer, Roda Verheyen, gave talks as part of the same climate change seminar series in Hamburg, albeit on different dates and different topics. Consequently, the defendant's lawyers expressed their worry that Otto and Verheyen "are part of the same interest and contact network."[14] The lawyers concluded that due to her public statements expressing interest in climate litigation and her association with "activist lawyers," Otto was biased in favor of the plaintiff and would be unable to examine the evidentiary questions from a neutral point of view if the court appointed her as an expert witness. Instead, they suggested that the judges appoint Judith Curry, a climatologist who has publicly questioned the scientific consensus on the anthropogenic contribution to global warming (Petersen et al. 2019; Nuccitelli 2019). Ultimately, the court did not choose any of the people suggested by the lawyers and found its own experts.

Contemporary judicial discussions about climate change echo legal disputes against tobacco companies decades earlier (Oreskes and Conway 2010). Science may play a similarly crucial role in litigation against fossil fuel companies (Geiling 2019). RWE's efforts to question the validity of climate research and attack scientists' credibility reflect a broader trend: Researchers have examined fossil fuel companies' public communications about global warming by comparing them to the tobacco industry's efforts to downplay the harms of smoking in previous decades. Naomi Oreskes and Erik Conway (2010) trace how major US industries have funded research that raised public doubt about the dangers of smoking and global warming. This research questioned the scientific consensus on those issues, highlighting underlying uncertainties. Notably, tobacco companies used such contrarian science as evidence in litigation over the harms of smoking. In addition, companies engaged in ad hominem attacks, questioning the credibility of scientists providing evidence on behalf of plaintiffs (Oreskes and Conway 2010). Major corporate polluters have deployed similar strategies to address climate change. Historical research shows that the US fossil fuel industry has promoted disinformation on global warming since the 1980s, despite the fact that industry representatives were briefed about the dangers of global warming at the time (Franta 2021). The industry has provided significant funding to US organizations that publicly question the scientific consensus on climate change and lobby against government climate action (Brulle 2014). Beyond outright climate change denial or skepticism, industry and other actors seeking to limit climate action have increasingly promoted discourses of "climate delay." According to a study by William Lamb and colleagues (2020), this strategy involves redirecting responsibility for climate change away from major polluters, pushing for nontransformational solutions, highlighting the downsides of climate policies, and surrendering to climate change as inevitable.

Climate science is often produced in response to policy makers' demands, public concerns, and researchers' own worries about global warming. To argue that this relationality of climate change knowledge makes it biased may lead to the absurd conclusion that all climate science is illegitimate for use in a legal framework. At worst, it means that the entire discipline of climate science is biased, echoing arguments of climate change deniers (Cann and Raymond 2018). In a similar vein, one might argue that cancer researchers are biased if they want to stop people from dying of cancer. RWE's lawyers contended that climate scientists who publicly expressed interest in climate change and the responsibility of corporations were unsuitable as expert witnesses in court. Following this line of argumentation, insights from peer-reviewed scientific publications cannot be trusted if the authors are suspected to have critical opinions about fossil fuel companies. Coauthorship and academic referencing are a sign of bias. The lawyers set high evidentiary hurdles to determine which scientific expertise should be accepted by the court: It may be difficult to find a climate scientist who has not expressed concern about global warming and has never been associated with or cited a scientist who believes that fossil fuel companies might have a responsibility to address climate change. It may be fruitful to avoid the naïve conception that science can—or even should—be entirely disinterested. The production of climate change knowledge is intrinsically entwined with people's concerns about devastating impacts felt around the world.

Entangled Responsibilities

Climate models and attribution science provide valuable knowledge about the processes and impacts of climate change. They involve inherent uncertainties, including limited knowledge about particular inputs and the physical randomness of atmospheric processes (Hulme 2010a). Attribution science is pertinent to legal and political disputes about who should take responsibility for climate change, particularly for negotiations about Loss and Damage at the United Nations (Burger et al. 2020). These discussions revolve around the irreversible impacts of climate change such as glacial retreat and sea level rise. Since the 1990s, vulnerable countries such as low-lying island nations have demanded that those who caused the most damage should pay reparations, but Global North countries have refused to accept any kind of formal responsibility. Attribution research can provide moral support to developing countries that have filed claims for compensation (Toussaint 2021). However, even if attribution science provides key evidence in climate litigation, it does not answer normative questions about which impacts should be prioritized and who should take responsibility (Hulme 2014).

Academic and legal debates continue about whether climate science holds up to judicial evidentiary standards. In their defense against Saúl Luciano Lliuya's claim, RWE's lawyers argued that climate models are entirely unsuitable for establishing causal liability. This assertion echoes discussions in other climate change lawsuits. In *Juliana v. United States*, a case brought by youth plaintiffs over the government's responsibility to address climate change,[15] the US government's lawyers attacked the reliability of attribution evidence and argued that confounding factors such as land use change and economic circumstances could not be disentangled from anthropogenic climate change impacts (Burger et al. 2020, 185).

These claims reflect critiques from social scientists who point to the reductive nature of climate modeling. At this point, a reminder may be helpful for politicians and lawyers that any distinction between "humanity" and "nature" is socially constructed (Ingold 2000). As attribution science provides an ever-more precise and detailed—albeit imperfect—representation of climatic processes, it offers a differentiated understanding of causality in the contemporary age. This approach to climate change moves beyond the idea that nature and humanity occupy separate spheres of existence, and toward a new understanding of planetary entanglements and human responsibilities (Hastrup 2013). Scientific advancements allow us to conceptualize the climate as a set of relationships in which we are all involved. Climate science allows for claims that we are all potential neighbors, whether in an individualized or universal sense. Humans are part of the nature that climate science describes (Knox 2015). Scientific knowledge points to the entanglement between people and environment. It opens novel possibilities for making claims about accountability, highlighting specific relations in that entanglement. Climate models and attribution science undoubtedly have limitations, but they are the best epistemological tools available for understanding climate change, developing policy responses, and dealing with questions of responsibility and justice.

7

Measuring Palcacocha

On a crisp Andean morning in 2017 under a clear sky, the sun began to rise above Lake Palcacocha. Lying beneath heavy woolen blankets, Martín Amaru[1] awoke in his stone shack above the lake to the sound of radio static. It was around six a.m., and Amaru had drifted off after his last radio message to the city one hour earlier to update authorities about the lake's status. Opening the wooden door, Amaru peered out of his shack. The sun was rising over a magnificently blue lake and sparkling white glaciers. According to a 2009 measurement, the lake had grown dramatically, and authorities feared that it posed a risk to the valley below: If a piece of glacial ice or part of the natural moraine dam fell into the lake, it could produce a devastating flood wave that could destroy the city of Huaraz. To counter this threat, the government had installed siphons to reduce the lake level and planned to build a larger dam and drainage system to replace two smaller concrete dams built in the 1970s. Amaru was part of a small group of men employed by the regional government to watch over the lake and keep the

FIGURE 7.1. The siphons at Lake Palcacocha. (Photo by author)

siphons running. In the case of a flood, their task was to warn the authorities via radio and allow for an evacuation.

I had spent the night in another shack by the lake and now joined Amaru on his morning routine. Descending along a steep slope, he skipped across rocks like a mountain goat. From the base of the slope, I followed him on a ten-minute walk over hilly ground to the lake's edge. Standing on the concrete dam, he surveyed Palcacocha as the first rays of sun hit his face. The wind was causing a light ripple on the lake's surface. Every few minutes, a crashing sound from the glaciers disturbed the quiet morning. These were only small avalanches.

Walking up to Palcacocha's edge, Amaru inspected the black siphons that stuck out of the water and went through a tunnel inside the dam (figure 7.1). He knocked on some of them with his fist and was satisfied with the quiet thud. They were all in working order, carrying water from the lake into the river below. Walking along the lake's right edge, he inspected a measuring stick in the water, making a mental note that the water had sunk by half a centimeter since the previous evening (figure 7.2). Morning inspection—done.

To determine whether Palcacocha posed a "serious hazard" of flooding, the court in Hamm, Germany, needed solid facts. Someone had to measure the lake and determine whether a flood was likely. Amaru took measurements every day and shared them with government authorities. Officials drew on

FIGURE 7.2. The measuring stick at Lake Palcacocha. (Photo by author)

his data to argue that the lake was safe or that it required enhanced safety works. Officials with other state agencies and scientific researchers took alternative measurements with different tools, giving rise to rival figures and facts for Palcacocha. The judges required accurate measurements to determine whether Palcacocha posed a flood risk. Someone had to gather data to support scientific facts about Palcacocha's situation. Understandings of *accuracy* vary depending on how the term is defined; each regime of measurement defines its own standards of accuracy. Measurement can provide legitimate knowledge if people trust its procedures and precision (Porter 1995).

On their own, the figures that Amaru gathered had little meaning beyond their immediate context. They were numbers on a measuring stick in the water. If conducted effectively, measurement allows for localized practices to generate knowledge that meets a universal standard; separating knowledge from its context of origin generates "objective" results (Porter 1995, 22). Turning measurement data into scientific facts involves a process of abstraction. Abstraction means transforming a phenomenon to make it understandable to a wider audience. The scientific process involves long chains of abstraction and transformation. This makes the phenomenon increasingly commensurable with other phenomena according to universal standards, but obscures the circumstances from which it arises (Latour 1999). The lake was transformed from water hitting a measuring stick to a number in a notebook. Later, scientists and government officials used that number to claim that the lake level had dropped. In a further abstraction, they argued that the risk of flooding was now lower. Authoritative statements about Palcacocha involved taking measurements at the lake and abstracting the data. This gave rise to scientific facts.

Arriving back at the shack above the lake, Amaru started boiling water over a wood-burning stove for his breakfast of porridge and tea made with fresh mountain herbs. He checked his wristwatch and saw that it was almost seven a.m.—time to radio the city. He walked over into his room, turned on the two-way radio, and spoke into the mouthpiece in Spanish.

"Base one to base two."

Radio static.

"Base one to base two."

Radio static. A few moments later, finally the response: "Base two to base one. I hear you. Please go ahead."

"Copy, base two." Amaru cleared his throat.

"Here the report for seven a.m. Weather: overcast. Wind: normal. Rain: none. Snow: none. Equipment: eight siphons operating. Water level:

sinking. Level decreased 0.5 centimeters. Avalanches: minimal. Waves: moderate. Mudslides: none. Everything is operating; no news, base two. Please respond."

Radio static.

"Base one, please repeat lake level."

"Lake level decreased by 0.5 centimeters."

"Copy that, base one. Base two, out."

"Affirmative, base two. Thank you. Until next time."

Amaru took his job seriously. Day after day, he made the call every two hours. He and his colleagues took turns working through the night. They could hardly sleep, staying attentive to the lake and mountains. Sometimes nobody answered the call at the Regional Government Emergency Response Center in Huaraz, in which case the clerk had presumably fallen asleep at their post or stepped out for some food. In the event of a flood, the workers had to call down and warn the authorities via radio. They also had a satellite telephone and a list of cell phone numbers to call—the police chief, mayor, the governor. In the meantime, Amaru stuck to his post and kept watch over the lake.

Amaru's shack above Lake Palcacocha had a table that held the two-way radio, powered by car batteries that he recharged with solar panels on the roof. Next to the radio in his dark room lay a stack of lined notebooks, the type children use at school. Every two hours, he recorded data by hand based on his measurements: wind, rain, snow, numbers of siphons in operation, water level, number and severity of avalanches, waves, and mudslides. He measured the lake level twice a day. He and his colleagues transmitted these data to a government employee in Huaraz, who entered the numbers into a computer. The data were later compiled into internal government reports about the lake. Periodically, the government published press releases based on these figures, and officials made declarations to the public about Palcacocha. To counter people's fears about flooding, officials often claimed that the situation was under control: they were monitoring the lake and the risk was low.

Effective measurement relies on the standardization of instruments and processes. Measurements are comparable only if everyone measures in the same way (Porter 1995, 29). To the naked eye, Lake Palcacocha is a large mass of water surrounded by towering mountains topped by glaciers. Measuring the lake according to universal standards enables people to make abstract claims about its status. Through measurement, Palcacocha becomes more or less a threat. Amaru was one of many people gathering data at the lake:

Members of several government agencies took measurements and produced a confusing—and sometimes contradictory—collection of facts, statements, and publications. This ultimately led the judges in Hamm to seek independent advice to clear up the situation.

The judges' job was to determine the facts. Initially, the question of flood risk seemed straightforward to Luciano Lliuya and his legal team. After arguing that RWE contributed to global climate change and that climate change led to glacial retreat in Peru, the lawyers completed the causal chain by linking glacial retreat to a risk of flooding affecting Luciano Lliuya's house. Numerous studies in peer-reviewed scientific journals identified a high risk of flooding from Lake Palcacocha, located above the city of Huaraz. The glacial lake had already caused a devastating flood when its banks burst in 1941, and as glaciers had retreated in the previous decades, it grew to be larger in volume than ever before. Local authorities began implementing infrastructure measures to reduce flood risk in 2011. Nevertheless, RWE's lawyers argued throughout the legal proceedings that there was no serious risk of flooding or that, at the very least, the risk was not serious enough to warrant a legal claim. Discussions revolved around scientific practices for evaluating flood risk at Lake Palcacocha. As various scientists and government agencies conducted research at the lake, competing data and understandings of risk became the subject of legal dispute. Lawyers on each side drew on scientific facts that justified their arguments about glacial retreat and flood risk. These facts were key to establishing—or contesting—a neighborly relation between Luciano Lliuya and RWE.

This chapter examines the last link in the causal chain: that between glacial retreat and Luciano Lliuya's house. The court heard lengthy arguments about which facts were most trustworthy. I trace how factual claims about Lake Palcacocha emerged and entered the legal proceedings. I follow the process of fact production from measurement practices at the lake, to scientific publications, to legal briefs submitted to the court. The lake itself is in constant flux; its shape and volume shift from day to day. It tends to grow during the rainy season and shrink during the dry season. Every few weeks an avalanche stirs up waves. Scientific measurement practices capture a momentary picture of the lake. Based on these measurements, scientists, experts, and legal practitioners evaluate the risk of flooding. Risk evaluations are shaped by institutional understandings of danger and the normative interests of evaluators. I follow how facts about Palcacocha came into being and affected people's lives between Huaraz and the German courtroom.

Contested Measurements

Early one morning in 2018, I met Pedro Vasquez[2] in Huaraz to join him for his monthly visit to Lake Palcacocha. It was another sunny day as we took the Glacier Authority's pickup truck up bumpy dirt roads. The ride took around two hours. We talked politics; he said he was fed up with corrupt politicians who prevented Peru from developing. He came from the city and had studied engineering at the local university. Now in his late thirties, he had spent most of his working life with the Glacier Authority. Vasquez was part of the educated urban class of Peruvians who earn a comparatively good salary and live a comfortable life.

Vasquez was in charge of glacial lake monitoring at the Glacier Authority. He performed a bathymetric study at Palcacocha in 2016 that was presented as evidence in the German court. Bathymetry involves a precise measurement of lake volume using sonar technology. The result was a three-dimensional image of the lake and a precise assessment of its shape, volume, and location at the time of measurement (figure 7.3). The most recent bathymetric studies, conducted in 2009 and 2016, showed a volume of 17.3 million and 17.4 million cubic meters, respectively. However, Vasquez and his colleagues at the Glacier Authority took those measurements when the lake was at particularly high levels. After the 2009 measurement, authorities installed siphons and reduced the level by several meters. By 2016, when Vasquez took the next measurement, the lake had expanded significantly, but it had decreased by several meters when I last visited in September 2024. Despite the fluctuations, scientists argued that the risk of flooding remained high. Amaru, Vasquez, and others have taken less precise, visual measurements at regular intervals. Once a month, Vasquez visited Palcacocha to read the lake level.

We arrived at Palcacocha. The road ended by the siphons that continuously pumped water out of the lake into the river that fed Huaraz with potable water. Vasquez and I carried the measuring equipment—several heavy black-plastic boxes and a tripod—on the fifteen-minute walk up to the concrete dam on the lake. With a panoramic view of Palcacocha, Vasquez set up the tripod by the dam. Pointing to the black box I was carrying, he said, "Hand me the thing in that case." I set the box down on the dam and retrieved a small electronic device with a screen, little buttons, and an eyehole. As I handed it over and Vasquez attached it to the tripod, he explained, "This is a laser measurement device—a total station. It gives me an exact reading of the lake's surface level." He pointed it toward Palcacocha and activated the device. After punching some buttons and making a few more adjustments,

FIGURE 7.3. The 2016 Palcacocha bathymetry rendered in a two-dimensional model—each line represents a gradation in depth. (Cochachin Rapre and Salazar Checa 2016)

he was satisfied. "Now it's saved the measurement. Later in the office I can download that to my computer."

"So can you measure the lake volume or only the water level?" I asked.

"What I'm measuring here is the precise water level. I can see how high it is above sea level and how much it's changed since the last measurement. That shows us how the lake is developing—now we can see that it's dropping because we're entering the dry season. If I want to know the volume, I place this water level onto the diagram from the last bathymetric study I did two years ago. Since then, the level has decreased by around three meters, which should equate to about two or three million cubic meters. That means the lake now has a volume of fourteen to fifteen million cubic meters."

"But what if the lake has changed shape since the last bathymetric study? Then the model wouldn't be accurate anymore, right?"

"Yes, that's the problem. The lakebed tends to transform over time. You can see that if you compare the last two bathymetries from 2016 and 2009: the shape of the lake actually changed during that time, even if the volume was still similar. When I do the laser reading, I can only estimate the volume; it's not a precise measurement." We packed up the total station and tripod. Although the lake workers took analogue measurements with a measuring stick every day, Vasquez preferred his electronic readings.

Measurement involves underlying assumptions about what counts and what we can know about the world. At Palcacocha, measurement was contested in terms of accuracy. Vasquez argued that his measurements were more precise than others'. Quantitative figures appear to provide authoritative and objective facts, yet they often embody theoretical assumptions. Numbers seem to speak for themselves but can say much more than what they explicitly describe (Poovey 1998), as when officials understood a higher lake measurement to mean that the risk of flooding had increased. At Palcacocha, the lake worker Amaru measured daily changes to the lake's water level. He reported to officials at the regional government who regarded this information as important. Every month, the engineer Vasquez measured Palcacocha's surface level. This allowed him to monitor its regular depth changes during the dry season that lasts from May to September each year. At longer, irregular intervals, the Glacier Authority conducted bathymetric studies to determine the lake's precise volume and contour. Each of these measurements gave rise to alternative facts that later entered the legal sphere, in a large binder on Judge Rolf Meyer's table in the courtroom. Palcacocha measurements allowed Peruvian officials, international scientists, and German lawyers to make competing factual claims about whether the lake posed a flood risk. Whoever wants to establish the facts about Palcacocha must decide which types of measurement should count.

Palcacocha Becomes a Scientific Fact

Lake Palcacocha has appeared in international scientific journals as a prime example of glacial lake danger since the mid-2000s (Vilímek et al. 2005; Somos-Valenzuela et al. 2016; Huggel et al. 2020). Scientists from countries including the United States, Switzerland, and the Czech Republic have visited the lake and produced scientific assessments about flood hazard. In addressing Palcacocha, they had to choose from multiple contradictory lake measurements. These measurements became a key dispute in the lawsuit over RWE's contribution to flood risk at Palcacocha.

The construction of facts is a collective process. Facts receive legitimacy when scientists or other authoritative figures argue that a statement is true (Latour 1987). When researchers at the Center for Research in Water Resources[3] at the University of Texas (UT) at Austin began modeling a potential flood disaster at Palcacocha, they had to choose among several lake volume assessments. Based on the lake workers' daily readings and measurements by various state agencies, members of different Peruvian government authorities had made numerous statements pointing to different figures. The UT scientists settled on the figure that appeared most trustworthy and precise: the 2009 bathymetric study indicating a water volume of 17.3 million cubic meters. This measurement may have appeared most accurate because it offered a three-dimensional picture of the lake. The spatial detail allowed researchers to model a potential flooding event. Based on this model, they simulated avalanches, subsequent lake behavior, moraine erosion, and downstream flooding (Somos-Valenzuela et al. 2016). As they reproduced this figure and published articles in renowned scientific journals, 17.3 million cubic meters appeared to be one of the more stable scientific facts for Palcacocha.

When the Peruvian press picked up on the UT studies and 17.3 million cubic meters figure, officials of the regional government in Huaraz began to worry that this volume fact—which pointed to a significant flood hazard—would alarm the population. This news could threaten the tourism industry, a vital economic sector in the region. Regional government officials publicly criticized the University of Texas study, arguing that the water volume had decreased to 14 million cubic meters and the lake was now safer. For outside observers, this destabilized the 17.3 million figure and claims about flood hazard. With no stable facts in sight, the judges on Luciano Lliuya's lawsuit had little choice but to seek additional expert advice. Usually, scientific fact production is an ongoing process marked by disagreement. Fact making is particularly difficult in the case of Palcacocha: The lake rises and falls with the seasons while the glaciers that feed it continue to recede, altering its shape and the likelihood of overflow. Measurement practices can describe the lake's volume only at a single moment. This snapshot description obscures the lake's temporal variability to government officials and legal practitioners who seek to determine whether Palcacocha poses a danger based on "accurate" measurements.

Facts Have Moral Contours

Broader concerns often shape the production and communication of facts (Latour 2004). Social context affects how people weigh factual claims. Wider issues can affect the process of fact production by legitimizing certain types of knowledge and excluding others. Fact making at Palcacocha is political: Knowledge shapes the priorities and scope of political action. An objective factual foundation can lend credence to a political project. If the facts are clear, solutions may seem obvious (Porter 1995, 7). At the same time, people's political concerns affect which knowledge they consider to be relevant and legitimate. When officials made factual statements about Palcacocha, their declarations were often linked to social and institutional interests.

In discussions about Palcacocha, some officials drew on measurements pointing to decreased water volume to calm public worries about flood hazard. At the same time, scientific claims about imminent danger justified government authorities' efforts to reduce the threat of disaster with flood prevention infrastructure at Palcacocha.

In January 2011, the Peruvian president decreed a state of emergency for Palcacocha due to an imminent flood hazard.[4] The decree cited the bathymetry figure of 17.3 million cubic meters and called for immediate measures to reduce danger. As a result, regional government officials in Huaraz initiated a safety project to install siphons at Palcacocha with the aim of reducing the water volume and concurrent danger. When the Peruvian president encountered scientific facts pointing to a mortal flood hazard at Palcacocha, he swiftly implemented measures to reduce the danger and avoid loss of life.

Claims about Palcacocha occur in the context of wider concerns about climate change. The Palcacocha flood hazard is an emblematic climate change impact that even appeared in the IPCC's *Sixth Assessment Report* (IPCC 2022). In discussions of climate change, epistemology is linked to morality and politics: It defines what counts as knowledge, which knowledge matters, and how it should drive action (Callison 2014, 14). Factual claims about climate change are inherently connected to conceptions of what should be done about it. These conceptions relate to more than adaptation measures, or how specific impacts such as the rising glacial lake waters should be addressed: they can also support claims against those who are thought to be responsible for climate change. The Palcacocha facts not only justified an infrastructure project but also provided Luciano Lliuya and his legal team with a factual basis for a climate justice claim against RWE. Building on morally charged knowledge about flood risk at Palcacocha, the legal complaint sought to establish an ethical neighborly relation between the plaintiff and the defendant.

RWE's lawyers countered the legal and scientific arguments at all levels. In an April 2016 legal brief, they cited a year-old local news report from Peru based on a regional government official's declaration that siphoning had reduced the lake level from 17 million to 12 million cubic meters. "For this reason," the lawyers concluded, "the defendant denies that Lake Palcacocha currently poses an acute risk of flooding."[5] Unbeknownst to the lake workers, RWE's claim arose from their daily measurements at Palcacocha. Having neatly compiled them in their notebooks, the workers communicated the figures to the regional government. Based on those figures, an official declared the lake safe to a local journalist, allaying public worry about flood risk and concern that a project to build a new dam at Palcacocha was not moving forward. In an effort to avert legal responsibility for climate change, RWE's lawyers cited the official's claim.

Meanwhile, Pedro Vasquez of the Glacier Authority conducted a new measurement at Palcacocha. In early 2016, he hiked up to the lake with mules carrying his inflatable rubber boat. The road to Palcacocha was not built until later that year. Vasquez spent five cold days on the lake navigating back and forth across the surface in his boat with a sensor. Finally, he had compiled a complete electronic reading. Feeding this into his computer, he produced a three-dimensional image of Palcacocha and found that it held a volume of 17.4 million cubic meters. Based on these measurements and calculations, he filed a report with his boss. Luciano Lliuya solicited the report from the Glacier Authority. It appeared as evidence attached to a legal brief in September 2016 countering RWE's arguments. Luciano Lliuya's lawyers argued that the regional government's claim cited by RWE was dubious. According to the glaciologist Christian Huggel, who provided scientific advice to Luciano Lliuya's legal team, such statements from Peruvian government officials are often unreliable.

In dismissing the lawsuit in November 2016, the Essen State Court found that it was not possible to establish legal causation via global climate change. Luciano Lliuya appealed that decision, and the legal dispute moved to the Upper State Court in Hamm, where the claim found unexpected success. Meanwhile, local disputes over flood risk from Palcacocha continued in Huaraz. In March 2017, representatives of the Peruvian Comptroller's Office, a national oversight agency, visited Lake Palcacocha and concluded that the water level had continued to rise and the risk of flooding was high. They called on the regional government to quickly implement safety measures (La Contraloría General de la República 2017). National news outlets picked up the story that Palcacocha continued to pose a threat to the city of Huaraz (Urbina 2017). In Huaraz, authorities hastily organized a press conference during which they dismissed the comptroller's report as scaremongering. The head of the national Glacier Institute (formally the Instituto Nacional de Investigación en Glaciares y Ecosistemas de Montaña [National Research Institute for Glaciers and Mountain Ecosystems]) argued that the water level had decreased, though the lake still posed a "latent risk." He drew on measurements that Glacier Institute employees had conducted. RWE's lawyers found valuable fodder for argumentation in a press release published by the Glacier Institute (INAIGEM 2017). In an October 2017 legal brief, they cited this report to counter the argument that Palcacocha continued to pose an immediate threat.[6] Drawing on a new set of facts, they once again portrayed the lake as safe and under control. In 2018, the judges in the Upper State Court in Hamm appointed two scientific experts to provide an independent opinion on the matter of flood risk at Palcacocha.

In December 2021, RWE's lawyers introduced surprising new evidence. In a recent study (Kos et al. 2021), researchers had analyzed ice movement in the Palcaraju glacier above Palcacocha, using satellite radar data gathered between 2017 and 2020. The study found that while some smaller avalanches had been detected, "an imminent failure of a glacier instability was not evident during the observation period" (Kos et al. 2021, 16). Based on this conclusion, RWE's lawyers contended that "there are no indications for a concrete and immediate threat of a GLOF-event through an ice avalanche at Lake Palcacocha."[7] The study appeared to provide damning evidence, undermining the plaintiff's argument that he faced a significant risk of flooding from Lake Palcacocha. However, the study examined only ice movement and did not evaluate the stability of the underlying permafrost. Larger glacial lake outburst floods typically occur when the mountain rock below the glacier becomes unstable. The article appeared in the journal *Remote Sensing*, which is published by the Multidisciplinary Digital Publishing Institute. It is open access, which according to the journal's website meant the authors paid an Article Processing Charge of 2,500 Swiss francs.[8] The article's publication process was remarkably short: It was submitted on June 10, 2021, revised on July 6, and published on July 8. The journal attracted controversy in 2011 when it published a paper that called into doubt widely recognized scientific insights on the anthropogenic contribution to global warming,[9] leading to the resignation of the journal's editor. Furthermore, the circumstances of the study's production raise questions.

In their legal brief, RWE's lawyers stated that they did not commission the publication but that the company had previously commissioned the firm Terrasense Switzerland, which was led by the article's lead author, Andrew Kos, to analyze the radar data. One of the study's other authors is Florian Amann, faculty at RWTH Aachen who provided scientific advice to RWE during the legal process. Investigative journalists later found that RWE paid Terrasense €40,000 to conduct the analysis and that Florian Amann charged RWE €120 an hour to work as an expert on the case, with the money paid to his university (*SourceMaterial* 2022). In the acknowledgments section of their study, Kos and colleagues (2021) state that the underlying satellite data "were provided through a commercial data purchase" and that "permission to use the results for this publication was granted by the firm RWE AG." According to a media investigation, RWE paid Airbus €60,000 for the data and provided it to Terrasense (*SourceMaterial* 2022). The authors make no further mention of RWE, nor do they acknowledge the ongoing lawsuit. They claim that "this research received no external funding," and they "declare no conflict of interest."

The journal's "Instructions for Authors" require that "all sources of funding of the study should be disclosed."[10] Authors must "declare any personal circumstances or interest that may be perceived as influencing the representation or interpretation of reported research results." Such circumstances include the role of funders "in the choice of research project" as well as "the collection, analyses or interpretation of data." The guidelines further state that "projects funded by industry must pay special attention to the full declaration of funder involvement" and that "*Remote Sensing* does not publish studies funded partially or fully by the tobacco industry." Potential conflicts of interests, the journal advises, may arise from "consultancies" and "paid expert testimonies." In response to a media inquiry, coauthor Florian Amann said that they "wrote the paper unpaid, independently without reference to the court case and uninfluenced by any third parties" (*SourceMaterial* 2022).

Facts arrive in the courtroom via a complex process of measurement, scientific fact making, and legal storytelling. Measurement practices shape our knowledge of the world. Divergent measurements at Palcacocha gave rise to contradictory assessments about the lake's safety. Scientific facts about Palcacocha have seen their validity disputed. Often, the disputes arose in relation to broader concerns about public safety or climate change. Facing an ongoing climatic threat to his livelihood, Luciano Lliuya took the matter to court. Lawyers on both sides of the legal process sought out convincing facts to strengthen competing responsibility narratives. Judges hold the power to turn scientific findings into legal facts. Based on these facts, the court in this case had to decide whether to establish a neighborly relation between Luciano Lliuya and RWE. Although legal knowledge is specific to individual cases, it can inform other courts' reasoning about causation and accountability.

IN THE *LUCIANO Lliuya v. RWE* trial, legal practitioners strategically deployed scientific evidence to make and unmake a neighborly relation between the plaintiff and defendant. Discussions revolved around the causal chain linking RWE's power plants, the global atmosphere, melting Andean glaciers, and Luciano Lliuya's house. While Luciano Lliuya's lawyers presented a quantifiable link between RWE's emissions and global warming, the defendant conceptualized these processes as too complex to capture scientifically. Lawyers argued about the use of climate models as legal evidence. They drew on competing lake measurements to evaluate the risk of flooding that Luciano Lliuya faced.

The suit was dismissed in 2025 after the court found that the flood risk to Luciano Lliuya's property was not sufficiently high to justify a legal claim. However, the court affirmed that, in principle, a causal link between major greenhouse gas emitters and specific climate harms can be established. In regard to the causal chain, the court found that it was conceivable to establish a "direct" connection between the defendant and plaintiff "since the processes set in motion by [RWE's] actions are almost linear and follow scientific laws."[11] This set an important legal precedent and confirmed that the plaintiff's legal approach was plausible—even if, in this instance, the evidence was not strong enough to satisfy the court.

Discussions about causation in the lawsuit centered on what kinds of relations can be established in times of climate change, which types of evidence and expertise produce those relations, and which norms should govern them. When lawyers and judges engage with climate change science, they must make sense of its underlying complexity and uncertainty within the legal framework. Causality is a key issue at stake in political and legal discussions concerning climate change. Arguments about causation and responsibility seek to construct politically charged socio-material relationships across scales of time, space, and social power. Neighborliness arises out of a causal link: Neighbors are those who can cause harm to each other, and neighborly relations are enacted through assertions about causal responsibility. Causality claims entangle communities, corporations, governments, and nonhuman persons in webs of responsibility.

Part III

MELTING GLACIERS PLAY POLITICS

On a dark evening in May 2017, a large avalanche crashed into Palcacocha. The deafening noise frightened Martín Amaru, who was keeping watch in his small stone shack above the lake. Employed by the regional government, Amaru spent weeks at a time at the isolated lake, often enduring freezing temperatures. In the moonlight, Amaru saw that the avalanche caused waves several meters in height. The water did not spill over the concrete dams, built to prevent a flood in the valley below. Amaru's job was to keep flood prevention infrastructure running at the lake. The waves washed ten of the siphons that continuously pumped water out of Palcacocha up onto the shore. Via his two-way radio, Amaru informed the authorities in Huaraz about the event: There would not be a disastrous flood tonight, but it was a close call. At five o'clock the following morning, Amaru still lay in bed as the sky beyond the mountains showed the first signs of illumination when another avalanche came roaring down. Again, the dams held, and Amaru dutifully informed the city officials.

In Huaraz, Amaru's calls triggered a significant commotion. After initial news reports, several state officials, seeking to avoid causing panic, declared publicly that the lake was safe. The existing flood prevention infrastructure had prevented disaster. Public discussions flared up in a familiar cycle: For years, scientists had warned of a significant risk of flooding from Palcacocha. According to hazard maps that officials had hung up in shops and restaurants throughout the city, much of Huaraz faced destruction if a large wave came crashing over the banks. Confusion reigned as people argued over how high the risk was and how authorities should address the issue. Two concrete dams had been built on the lake in the 1970s when it was much smaller. For several years, public officials had spoken of building a new dam and drainage system, along with an early-warning system against flooding, but both projects had yet to materialize. Many were also concerned as the river emerging from Palcacocha provided the city's water supply, arguing that draining the lake might leave the city with insufficient water.

Later I spoke to Eduardo Díaz, who comes from a nearby village and was the supervisor for the flood safety works at Palcacocha. He had spent most of his eighty-three years living and working in the high Andes. His skin was rugged from toiling in the fields. Now Díaz normally spent his days at the lake alongside Amaru. On one of my first visits to Palcacocha he explained how he maintained a close connection with the lake and surrounding mountains. I watched him perform a *pago*, a ritual offering to the landscape, which he did every few weeks to keep the mountains and lake happy. When the avalanches took place, Díaz had not visited the lake for several weeks due to problems in his village. I asked Díaz about the avalanches. He told me that the mountains were angry because he had left them hungry. Díaz soon returned to Palcacocha and continued performing *pagos*. There was no major avalanche for the rest of the year.

The lake holds many meanings for many people. For the inhabitants of Huaraz, it is a vital source of water, but many also worry about its potential as a mortal threat. State authorities approach Palcacocha as an object of techno-scientific intervention. Using scientific knowledge and engineering solutions, officials hope to contain the danger of flooding and benefit from the lake's waters. Those who worked at Palcacocha sought to implement a project that intervenes in the environment's flows, yet for them, this intervention would succeed only if they maintained a relationship of trust and reciprocity with the lake and surrounding mountains. The mountains and lake are neighbors of a different sort. For those who engage them as sentient beings, their power is unquestioned. Standing at the center of discussions about environmental change and global warming, Palcacocha has become a site of glacial politics.

The first section of this book traces how the lawsuit asserted a neighborly relation between Luciano Lliuya and RWE, framing climate politics as a neighborhood dispute. Now I examine who else is in the neighborhood. The following chapters explore the hidden stakes of Luciano Lliuya's claim against RWE. According to the plaintiff and his supporters, the lawsuit was part of a broader effort to address climate change by holding greenhouse gas emitters accountable. This attempt at far-reaching political action operated through a targeted legal intervention in the politics of glacial retreat in Peru, aiming to contribute to a flood prevention project at Palcacocha by drawing a foreign corporation into a neighborly relation. Contemporary glacial politics emerged amid tensions over how to understand and engage with glacial retreat in the Andes. In public discussions, people related glacial retreat and the growth of Palcacocha to looming flood risk, potential water scarcity, and the whims of powerful mountains. Building on scientific conceptions of risk and hazard reduction, flood prevention infrastructure emerged as the dominant approach in local glacial politics. This infrastructure brought together a variety of knowledges and socio-material relations. Sentient mountains do not play a formal part in the modern politics of glacial hazard management, yet they pop up in unexpected ways and shape people's engagement with the changing Andean environment. More may be at stake in glacial politics than initially meets the eye. These hidden interests reflect similar tensions in Luciano Lliuya's lawsuit. They highlight the fact that climate politics concerns more than modern institutions, holding the potential to recast the relations at stake in political discussions.

8

Glacial Politics

On a cloudy afternoon in Huaraz, Luciano Lliuya and I set off for Palcacocha. It was March 2017, and the rainy season was in full force. The skies often cleared up in the mornings, allowing for some sunshine as people left their houses for the day. A German journalist was visiting Huaraz to report on Luciano Lliuya's story. Driving in Luciano Lliuya's Toyota station wagon, we accompanied her to Palcacocha. Huaraz lies at just over 3,000 meters above sea level, high enough to leave most visitors out of breath from something as simple as walking up a flight of stairs. We took a dirt road from the city and drove upward along the Cojup River, which originates at Palcacocha and supplies Huaraz with water. In the ongoing rainy season, the road was bumpy and full of potholes. After an hour in the car, we arrived at a large metal gate that blocked the entrance to the Cojup Valley where a new road would take us up to the lake. Going through the gate, we also entered Huascarán National Park.

The first time I visited Palcacocha, in late 2014, there was no road from this point onward. It was a grueling six-hour hike up to the lake at 4,500 meters above sea level. Since then, authorities in Huaraz had commissioned a project to improve mobility for flood safety workers at Palcacocha. Over the course of several months, some seventy workers from nearby villages cut a road out of the mountain environment. While they moved the earth with tractors and removed the most unforgiving boulders with dynamite, much of the work involved brute human force with hand tools.

In Luciano Lliuya's car, we drove slowly along the rough, narrow road up the valley. The skies began to darken. Soon, the rain would pour. About halfway to the lake we slowed down as a pickup truck approached us. Stopping alongside the other vehicle, I saw Fernando Vilca,[1] an engineer from Huaraz who oversaw the flood safety project at Palcacocha. Vilca greeted me through the open car window. He had worked at the lake since engineering works began in 2011. Despite chaotic administration in the regional government and intermittent payments to him and the workers, he stuck to his post. The previous year, he had overseen construction of the new road. Several times each week, he inspected the lake and ongoing work. I asked him how things looked.

"The water level has risen slightly. There's been some strong sunshine that's made the glacier melt more," he explained. Díaz, the project's foreman, was keeping a close eye on the situation. "I have all the siphons running to decrease the water level," Vilca went on, "and I asked Díaz to perform a *pago* to keep the situation under control. We can't control the sunshine, but the *pago* can appease the mountains and help prevent avalanches."

We waved to Vilca and drove on. As an engineer, he was implementing a state-sponsored infrastructure project to reduce the risk of flooding. In his work, he relied on scientific standards of measurement and hazard assessment. Yet for him, scientific engineering was not sufficient to address the problem. Decades earlier, while working as a mountain-climbing guide in the Cordillera Blanca, he had come to understand the mountains and lakes as living beings that require people's respect. For him, appeals to the landscape in *pago* ceremonies were an essential aspect of safety works at Palcacocha.

This chapter examines how contemporary worries about flood hazard in the Peruvian Andes became a local, national, and global concern. I build on Carey's 2010 environmental history of the Cordillera Blanca, which traces authorities' engagement with the changing landscape from the 1940s to the early 2000s. Historically, glacial politics in the Cordillera Blanca emerged over fears of glacial lake outburst floods following several notable disasters. Government officials developed techno-scientific standards to determine which glacial lakes in the region posed a flood hazard. By this process they identified numerous lakes as potential sources of disaster requiring infrastructural remediation. Government agencies oversaw the construction of glacial lake dams from the 1950s through the 1980s. Barriers put in place at Palcacocha during this time would resist heavy waves and prevent disaster decades later when glacial retreat became a topic of international concern. Historical standards for analyzing the Andean environment, focusing on potential flood hazard, have thus shaped the stakes of glacial politics to this day.

Infrastructure offers a fruitful site for studying how politics is enacted through techno-scientific interventions in the landscape (Appel et al. 2018). From the perspective of governing authorities, infrastructures allow for systematic control of environmental variability (Edwards 2003). They embody assumptions about how socio-material relations are and should be ordered, assumptions often reflected in technical standards that foreground certain types of knowledge while excluding other perspectives (Star and Lampland 2009). At Palcacocha, the standards in play in present and historical infrastructure projects point to different knowledges at stake in the politics of glacial retreat and, more broadly, in neighborly disputes about climate change. On a daily basis, numerous systems of standardization guide behavior and social relations. Standards may appear politically neutral, such as the measurement standards that Vilca and his workers applied to gauge the water level at Palcacocha. Susan Star and Martha Lampland (2009) argue that standards nevertheless embody ethics and values. The standardized process of water-level measurement renders a representation of Lake Palcacocha along fixed lines. It may reveal that the water level has increased by five centimeters. While this representation may appear strictly technical, the applied standard excludes other possible conceptualizations: Has the lake shifted horizontally? Has its mood changed? Applying a particular standard involves a moral choice as it makes other perspectives invisible. The process of standardization constrains a phenomenon within a set of standardized dimensions.

In infrastructural systems, different standards are often nested within each other. My conversation with Vilca on the road to Palcacocha revealed multiple standards at play, each seemingly linked to distinct forms of knowledge: While he relied on scientific and engineering standards to measure the water level and operate the siphons, he drew on Andean standards of engagement with a sentient environment to appease the mountains above Palcacocha and prevent a dangerous avalanche. Those standards were inherently entangled.

Scars in the Landscape

Continuing our drive to Palcacocha, we arrived at the workers' camp below the lake. Several stone huts stood in a small compound surrounded by a low stone wall that kept out nearby grazing cattle. We emerged from the car below dark clouds. It was late afternoon. Smoke rising from one of the houses indicated that the workers were cooking. Díaz greeted us as we entered the camp. He had arranged for us to stay the night.

Díaz, an old man, peered at us from under a wide-brimmed hat as we laid down our bags in one of the huts (figure 8.1). Díaz was born in a village just above Huaraz. The mountain waters irrigated his fields of potato and corn. He gave us a warm smile that showed a handful of teeth. Speaking Spanish with Quechua intonations, he told us to follow him to the lake.

From the camp emerged a steep road leading upward. Centuries before, expanding glaciers had carved a moraine into the landscape—a large mound of earth and rock that now towered over the houses. When the glaciers later began to recede, Lake Palcacocha grew behind the moraine, which formed a natural dam.

Palcacocha first came to widespread public attention when that moraine broke in 1941, causing a violent flood. Early in the morning of December 13, around 12 million cubic meters of water rushed out of the lake and down the valley, ripping away boulders, trees, and earth. Farther downstream, the flood swept up a smaller lake, Jircacocha (Carey 2010). It was just becoming light in Huaraz, one eyewitness told me, as masses of water and debris crashed through the city. He was a child then and was getting ready for school that morning. From the window of his family's house, which lay far enough outside the city center to escape the disaster, he saw the waves knocking over trees "as if they were toothpicks." The flood destroyed almost half of Huaraz and left around eighteen hundred people dead (Wegner 2014).

Walking with Díaz up the path to Palcacocha, we came through the break in the massive moraine. Its walls towered around us on either side. Feeling the altitude of 4,500 meters, I advanced slowly to catch my breath. More agile on his feet than me at this altitude, Díaz climbed on ahead to meet a small group of workers who were completing the last stretch of the road. It was tough work in this rocky mountain environment. With shovels and pickaxes, the laborers slowly cleared a way up the mountain. Dressed in orange jackets and red construction helmets, they chatted in Quechua as the sky grew darker. Grabbing a pickax, I quickly tired after helping them break apart a small boulder.

FIGURE 8.1. Eduardo Díaz at Palcacocha. (Photo: Alexander Luna)

In the decades following the 1941 disaster, Peruvian government authorities began conducting research and building infrastructure to address glacial lake safety. They put together a glacial lake inventory and organized numerous engineering projects to drain and dam dangerous glacial lakes. Through these measures, the Peruvian state assumed responsibility for disaster prevention while drawing the high-altitude environment and its inhabitants into the realm of government. These efforts were embedded in scientific measurement practices that rendered lakes as sources of potential disaster (Carey 2010).

Glacial lake flood hazard became even more urgent for Peruvian authorities when it threatened the country's industrial development: In 1950 a flood at Lake Jankarurish in the Los Cedros valley, north of Huaraz, damaged a major hydroelectric facility downstream. The following year, the Peruvian president established the Lakes Commission, a state agency tasked with analyzing glaciers and lakes for potential danger and implementing measures to prevent outburst floods. The newly appointed officials faced the significant challenge of inventing a classification system for glacial lake safety and developing engineering techniques to reduce danger (Carey 2010, 84).

As a first step, the Lakes Commission sought to compile an inventory of glacial lakes in the Cordillera Blanca. Officials did not know how many lakes there were or where dangerous lakes were located. Avoiding reliance on local villagers or field research to gather information, Lakes Commission experts used aerial photographs to identify potentially dangerous lakes, visiting the lakes only on exceptional occasions. Rather than using existing Quechua names, they gave each lake a number. This reflected an effort to apply a universalizing scientific standard, while also extending the state's reach to remote areas and communities (Carey 2010, 85).

Over time, government reports began to include local names for lakes and mountains, leading to a multilingual hybridization (Carey 2010, 86). Today, people speak of "laguna Palcacocha": *laguna* means "lake" in Spanish; *cocha* means "lake" and *palca* means "bifurcation" in Quechua, the latter word referring to the shape of the mountain behind the lake. A handful of lakes are now commonly known by their numbers, such as Lake 69 and Lake 513. This points to an imbrication of multiple naming standards that has shifted over time and reflects different knowledges at stake in discussions about flood risk.

Officials built on scientific conceptions of quantification and flood hazard identification to establish a set of standards for determining which lakes were dangerous and required intervention. In 1952, the Lakes Commission recruited a geologist from the University of California at Berkeley to assist in setting parameters for recognizing potential danger. With his scientific input, authorities identified characteristics linked to lake stability: Was the lake in direct contact with the glacier? How stable was the lake's natural dam? Based on this new classification system, authorities identified thirty-five dangerous lakes, of which twenty-five required immediate intervention. This system forms the basis for evaluating lake hazard in Peru to this day and has been applied as a standard for examining glacial lakes around the world (Carey 2010, 92). Explicitly designed to address flood risk, these standards might appear purely technical. Nevertheless, applying them in preference to other modes of understanding involves an underlying moral choice in that they take little account of local people's conceptions of the lakes or other concerns such as water supply.

Building on their mapping of the region's lakes, and employing scientific engineering standards, authorities began implementing infrastructure projects in the 1950s to prevent glacial lake disasters. This work usually involved partially draining lakes identified as dangerous and building dams to prevent them from breaking out. In a first step, workers constructed open canals or drainage tunnels to lower the water level. Subsequently, they built earthen or cement dams that engineers designed to withstand flood waves caused by avalanches falling into the lakes. At the base of concrete dams, workers usually installed a drainage canal to allow for a constant outflow, preventing the water level from rising above that point (Carey 2010, 93). In the 1970s, authorities completed two concrete dams at Palcacocha, one of which includes a drainage canal. The dams stand around seven meters above the lake's water level, meaning they can contain flood waves up to seven meters in height if the dams remain stable (Portocarrero Rodríguez 2014).

Back at Palcacocha with Luciano Lliuya and Díaz, we proceeded on the last stretch up to the lake. In front of us stood a steep slope that blocked our view of the glaciers. There we encountered the material traces of a recent infrastructure project: ten large plastic siphons expelling water from Palcacocha into the river that led downstream. As it slowly began to rain, we walked along the slippery siphons that led four hundred meters up and into the lake. Finally, we came to the concrete dam that had stood in place for more than forty years. After struggling to reach the top, we gained a panoramic view of Palcacocha and the surrounding mountains (figure 8.2).

The lake was a dark blue that stood out against the shining white glaciers towering above the water, almost two kilometers beyond our position at the lake's other end. We could hardly enjoy the picturesque sight as the rain

FIGURE 8.2. Palcacocha in February 2017 during an inspection by local and national authorities. (Photo by author)

began to pour down. Díaz rushed us to a little shelter beneath a sheet of corrugated iron beside the lake. We heard a thunderous noise in the distance—a small avalanche coming down.

After working at Palcacocha for several years, Díaz was familiar with the environment's power. Like Vilca the engineer, he was concerned not only with keeping the engineering works running but also maintaining a positive relationship with the Andean environment. While sentient mountains were an important consideration for Díaz and others working at the lake, they are conspicuously absent in government reports documenting past interventions at glacial lakes. Relying primarily on scientific forms of knowledge, these reports do not account for the mountains as actors, leaving them mute in the government-sanctioned historical record. From the perspective of state officials, these local engagements are simply not relevant for flood risk assessments. Oral histories compiled in the region, on the other hand, provide extensive stories about powerful mountains and lakes (Yauri Montero 2000). For Díaz, reciprocal engagement with the landscape was necessary to successfully address the issue of flood risk.

Hiding under the shelter, with Díaz calm as ever, we watched the raindrops splash into the lake, which took on an ever-darker shade of blue as the clouds came down on us. By this time, we could hardly make out the glaciers beyond the lake. Luciano Lliuya stared into the misty distance. Because of his actions, Palcacocha had become the subject of an international legal dispute.

The Politics of Glacial Retreat

In early 2017, local, regional, and national officials from numerous government agencies gathered in a conference hall in Huaraz to discuss Palcacocha. After years of deliberations, little visible progress had been made in building a new dam and drainage system or in establishing the early-warning system that authorities had announced long ago. Following plans drawn up in 2011, after measurements pointed to a dangerous increase in the water level, authorities installed plastic siphons as a stopgap to reduce the lake's volume before construction began on a new dam. While the road to Palcacocha completed in 2016 by the regional government facilitated the transportation of materials, further safety measures did not appear to be on the immediate horizon.

The crowd in the conference hall reflected a wide array of institutional actors with a stake in Palcacocha. Representatives of the Ancash Regional Government and the Municipality of Huaraz chatted with officials from the

Glacier Authority, the Glacier Institute, Huascarán National Park, the Ministry of Agriculture, and numerous other state agencies. I sat quietly in the audience alongside members of the local press. Leading the meeting was a retired submarine commander with the Peruvian navy who now oversaw disaster risk prevention for the Peruvian government and was visiting from Lima. Sitting at a table in front of the seated crowd, the admiral exuded authority and decisiveness, his firm posture and staunch expression projecting an aura of military discipline. His powerful voice pierced the room.

"Palcacocha is an extraordinary beauty, but it can also cause terror." While the lake was an important natural resource, he explained, the authorities must work together to prevent another catastrophe like the 1941 flood. They should finally establish an early-warning system and needed to work toward constructing a new dam. The admiral would seek support from the Peruvian president and his cabinet, but everyone must cooperate.

After regional and local officials provided updates on the current situation at Palcacocha, tensions rose over how to balance flood risk and water supply demands. A representative from the Huascarán National Park pointed out that Huaraz depended on Palcacocha for its water supply. He urged the regional government to consider this issue as part of the lake safety project: They should also build hydrological storage infrastructure to ensure a water supply. "We need a more holistic approach to this project," he explained. An official from the Ministry of Agriculture quickly agreed, pointing out that water resource management was crucial.

This intervention left the admiral visibly annoyed. "The first priority is to reduce the risk. We should not delay the project any longer. Remember the 1941 disaster!" Speaking with militaristic authority, he declared that saving lives was the most important issue. Anything else could come later.

Many people in the region, particularly in rural areas, are afraid of water scarcity. They depend on glaciers for their water supply. Glacial retreat is visible year after year, and people are worried that they will no longer have sufficient water for drinking and irrigation. From a scientific perspective, glaciers act as water storage devices. They melt slowly throughout the year, giving life to rivers and streams. When it snows, they rebuild mass. Because of accelerated retreat in recent decades, glacial ice is melting faster in many places than it can reaccumulate. Entire glaciers have already disappeared.

Whereas public authorities have focused on the issue of flood risk at Palcacocha, many farmers are much more concerned about water scarcity. "Throwing away the water is bad," one rural community leader told me in an interview. "I don't have a problem with the siphons, and it's OK to implement safety works, but they shouldn't throw away the water. They should build a reservoir." After all, Huaraz and the entire area depended on Palcacocha's waters. He had little trust in officials' statements about flood risk: "The lake won't break out. For many years they've talked about an emergency, but nothing happens," he argued. "I don't see any risk for the city of Huaraz." Another community leader told me that the authorities had made up the issue of flood risk for the purposes of corruption: "They only talk about that so they can implement projects and steal money."

Lake Palcacocha has emerged as a site of glacial politics. Government officials, citizens, and scientists wrangle over how state institutions should engage with the changing Andean environment. There is fundamental disagreement over how glacial retreat will affect social life, who can provide legitimate expertise, and what issues should take priority in the political sphere. As the political theorist Chantal Mouffe (2005) has observed, politics involves practices and institutions that aim to calm social antagonism and create order. Glacial politics in the Cordillera Blanca revolves around how people should understand and engage with glacial retreat. In this context, political claims involve efforts to expand the stakes of glacial politics—that is, to broaden people's understanding of what issues related to glacial retreat deserve public attention. Local political disputes set the stage for Luciano Lliuya's lawsuit against RWE but were often hidden in the case's public narration. Different ways of knowing the Andean environment defined the potential stakes of glacial politics and gave rise to social claims.

Rising Waters, Rising Concerns

After the 1941 outburst flood from Palcacocha that devastated Huaraz, the lake was almost empty. A measurement showed 0.5 million cubic meters of water. In 1974, the Glacier Authority completed two concrete dams at Palcacocha to prevent a future disaster (Carey 2010). After a mudslide in 2003 caused a minor outburst flood, measurement showed that the water level had grown to 3.8 million cubic meters. The flood partially eroded the lake's secondary dam. It did not reach Huaraz but damaged sanitation infrastructure and left the city without running water for a week. A subsequent measurement in 2009 showed 17.3 million cubic meters (Portocarrero Rodríguez 2014). As the glaciers behind Palcacocha retreated, the lake had grown by a factor of thirty-four.

Following these developments, Palcacocha and the Cordillera Blanca became a focal point for international scientific research on glacial retreat and GLOF hazard.[2] Cordillera Blanca glaciers are close to urban centers with established transport links, making them easier to access for research visits than other areas prone to GLOF, such as the Himalayas. The Cordillera Blanca offered scientists a laboratory for studying the impacts of glacial retreat as climate change rose on the international political agenda. With the precedent of a past disaster, Palcacocha is a particularly valuable site for studying the potential processes and impacts of an outburst flood. With support from European and North American research funding, scientific research came to shape the politics of glacial retreat in Huaraz.

The study by scientists from the University of Texas at Austin modeled how a GLOF at Palcacocha would develop in different scenarios. Simulating possible flood paths for avalanches of different magnitudes, the researchers found that an ensuing flood would take around one hour and twenty minutes to reach the city. Modeling also showed that the areas affected by the flood could be significantly reduced if authorities decreased the lake's water level. The authors explicitly recommended that authorities in Huaraz install an early-warning system at Palcacocha, enabling them to evacuate the population in the case of a flood, and consider lowering the water level. The authors contended that flood risk should be a priority for both science and glacial politics: "There is consensus among local authorities, scientists and specialists that Lake Palcacocha represents a GLOF hazard with potentially high destructive impact on Huaraz, and this consensus has been validated by the modeling results presented in this paper" (Somos-Valenzuela et al. 2016, 2538).

A visual representation of UT study's model soon became ubiquitous in Huaraz, as it was printed on posters hung in restaurants and public buildings throughout the city (figure 8.3). A team of scientists from the University of Zurich with financial backing from the Swiss government teamed up with the NGO CARE and local government authorities to produce hazard maps for Huaraz. This project divided the city into red, orange, yellow, and green areas, indicating varying degrees of flood risk. Ominously, the map designated certain areas as corpse collection centers.

For public officials, the map became a key point of reference for warning the population about flood risk. The flood model established a useful framework for understanding how glacial retreat could affect the local population. Authorities said that around 50,000 people lived in potentially affected areas. In an interview, one official worried that many did not take the risk seriously. Even if authorities initiated an evacuation process, he feared that 20,000 people would die.

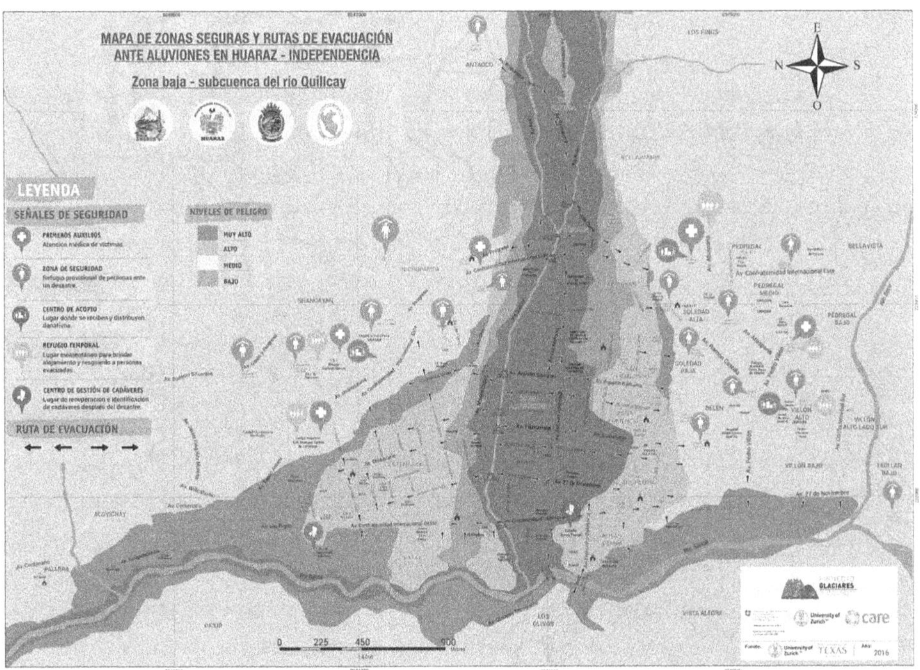

FIGURE 8.3. The flood hazard map of Huaraz. (Courtesy of Proyecto Glaciares)

With their flood-modeling scenarios for Palcacocha, scientists provided significant insights that affected political deliberations about glacial retreat in Huaraz. The hazard map lent epistemological credence to officials and members of the public in their calls for quicker implementation of flood risk infrastructure at Palcacocha. The need for infrastructure remained unquestioned in discussions among governing authorities and state agencies. They disagreed only about what type of infrastructure was needed and which types of knowledge should take priority in political decision making.

Some residents I spoke to, particularly in rural areas, criticized the overarching political concern with flood risk. They worried that this preoccupation drew attention away from water scarcity, which they regarded as a more existential threat. Inadvertently, the flood model set the stage for local glacial politics to address glacial retreat primarily in terms of flood hazard. Other scientific studies have focused on glacial retreat and water availability in the region (Bury et al. 2013; Drenkhan et al. 2015). Drawing on approaches in critical human geography, some have developed hydrosocial models that account for the entanglement of social and environmental change in the context of glacial retreat (Carey et al. 2014; Mark et al. 2017). These studies, however, have achieved less prominence in political discussions about glacial retreat in Huaraz than those focusing on flood risk. More recent research has proposed the implementation of multipurpose water management projects that address both flood risk and water supply (Drenkhan et al. 2019). Drawing on the latter approach, some officials have argued that the project to reduce GLOF hazard at Palcacocha should include water storage and irrigation infrastructure downstream, thereby addressing longer-term worries about water scarcity.

Political Fragmentation in Times of Global Warming

Shaped by scientific research about climate change impacts, glacial politics in the Cordillera Blanca revolves around how authorities should balance different knowledges and social needs and which infrastructures are most appropriate for addressing the problems at hand. As I discuss in part 2, in the legal context, scientific knowledge shapes how environmental change is understood and which solutions are potentially viable. In a fragmented political landscape, progress on infrastructural measures has been painfully slow since Palcacocha came to renewed public attention in 2009.

Decades ago, the Glacier Authority in Huaraz commanded significant resources to implement engineering projects to abate glacial lake hazard. But in the 1990s, the Peruvian government under the authoritarian president Alberto Fujimori implemented neoliberal reforms that fundamentally reshaped relations between state institutions, citizens, and the environment. The Glacier Authority in Huaraz, which had overseen infrastructural interventions at dozens of glacial lakes since the 1950s, was part of the state-owned regional electricity company. When that company was privatized and sold to US investors in 1997, the new owners shut down the agency. Local residents and glacier experts denounced this move, lamenting that authorities had rescinded their responsibility to keep the populace safe. The government reestablished a small glacier-monitoring office in 2001, but it no longer had the responsibility to plan and build large-scale lake safety infrastructure. Neoliberal reforms increased disaster risk and climate change vulnerability in the Cordillera Blanca (Carey 2010).

After widespread privatization and scaling back of the national Peruvian state apparatus in the 1990s, a process of decentralization in the 2000s transferred competences to regional and local authorities. In practice, the division of responsibilities between different levels of government often remained ambiguous (Pinker and Harvey 2015). In Huaraz, several governmental authorities and rival state agencies competed for influence in political discussions about flood hazard. Formally, the Ancash Regional Government was responsible for lake hazard in the Cordillera Blanca. In Huaraz, the municipalities of Huaraz and Independencia were also involved, as the hazard zone cuts through both jurisdictions.[3] While the regional government received significant funding via transfers from the national government and mining industry payments—enough to implement a large-scale glacial lake safety project—progress was slow in building a new dam and drainage system at Palcacocha. Numerous state agencies and nonstate organizations were also involved in the discussions. This included the Glacier Authority, which had a large staff until the 1980s and, in its heyday, oversaw construction of numerous glacial lake dams. Now called the Glacier and Lake Evaluation and Monitoring Area (Área de Evaluación y Monitoreo de Glaciares y Lagunas), it was part of the National Water Authority and had a small staff engaged primarily in monitoring. Another agency involved in monitoring was the Glacier Institute (INAIGEM), founded in 2014. Several other state institutions were also part of discussions, including the national agency in charge of disaster prevention and the Civil Defense Institute (INDECI), which oversees disaster response. Representatives of the national government in Lima have also joined some meetings focused on flood hazard at Palcacocha. Finally, NGOs such as CARE and the Mountain Institute have provided scientific and logistical support to state authorities in order to address climatic hazards.

Between 2011 and 2012, the Peruvian president issued twelve consecutive decrees declaring a state of emergency at Palcacocha, mobilizing resources for immediate action. Officials of the Ancash Regional Government drew up a plan to reduce the water level at Palcacocha and build a new dam and drainage system. Workers put in place a provisional siphoning system at the lake and remained permanently present to warn of a possible disaster.

In the following years, the Ancash Regional Government became mired in chaos. In 2014, Governor César Álvarez was arrested and removed from office pending charges of corruption and arranging contract killings of political rivals. Waldo Ríos, who was elected governor later that year, was arrested in 2016 and later convicted of corruption. His successor, Enrique Vargas, was in office for less than a year before being arrested and convicted in 2017 for presenting a fake university degree when he ran for office. Juan Morillo was elected governor in 2018 and held office until 2020 when he was arrested on corruption charges. With every change in leadership at the regional government, the incoming governor appointed new people to higher offices. Despite being formally responsible for overseeing glacial lake hazard in the region, many of those in charge had little knowledge of or experience with the issue. Vilca, the engineer tasked with overseeing infrastructural works at Palcacocha, explained to me that corruption was endemic to the regional government. He was cynical. "The new dam project hasn't happened in all these years, mainly because officials don't have a good chance to skim something off the top. They only care about siphoning money into their own pockets."

Amid political fragmentation and competing knowledges, progress in implementing long-term flood mitigation infrastructure has been slow. Construction of a new dam has yet to begin as of late 2025, and scientists continue to warn of a high flood hazard for the residents of Huaraz. How do workers overseeing infrastructural works at the lake balance conflicting knowledges and political priorities?

Balancing Flood Hazard and Water Scarcity

As state officials and members of the public discussed the value and potential danger of Palcacocha's waters, the engineer Vilca sought to find a practical balance in his job overseeing infrastructural works at the lake. On a clear morning in June 2018, I accompanied him in a pickup truck on an inspection at Palcacocha. It was the dry season, and the sun burned down on thirsty pastures in the valley below the lake. On encountering another pickup, carrying Pedro Vasquez, the official from the Glacier Authority (see chapter 7), we stopped for a quick chat.

Vasquez and Vilca were both concerned about the low water level at Palcacocha. If it sank any further, there might not be enough water to supply Huaraz. Vilca suggested installing siphons at Perolcocha, another lake in the valley, which could contribute to the river flow. In the meantime, he had three out of ten siphons running at Palcacocha. "I'm trying to increase the lake level," he told Vasquez, "but it's still going down." Once we arrived at the lake, Vilca instructed Martín Amaru to reduce the flow to one siphon until the water level rose sufficiently.

Formally, officials in the regional government were in charge of regulating the water flow at Palcacocha. In practice, they lacked expertise and left Vilca in charge. "Reducing disaster risk is our principal objective," Vilca later explained to me, but he sought to balance that with the need to maintain a sufficient water supply. "It's difficult to handle these two situations, but both are very urgent." While the river that emerges from Palcacocha used to receive significant input from water springs along the valley, they have begun to dry up in recent years. As the river flow has decreased, particularly during the dry season, local officials approached Vilca and asked him to increase the outflow from Palcacocha.

Vilca began to regulate the water level, allowing it to increase during the rainy season so there would be a sufficient supply during the rest of the year. "When the water level goes up, the risk also rises, but this is an issue of primordial human necessities." He felt a significant responsibility resting on his shoulders. "We need water to survive, so maybe we have to permit a little bit of risk." Through improvisation and learning on the job, he sought to keep Huaraz safe and maintain a sufficient supply of water.

Flood infrastructure at Palcacocha is a site of glacial politics as people deliberate over how to engage with the changing environment. In a fragmented political context, lake workers have struggled to balance the different claims and knowledges at stake. Meanwhile, Peruvian glacial politics has become embroiled in global concerns about climate change and glacial retreat, as dramatically reflected in Luciano Lliuya's legal proceedings in Germany. Officials in Huaraz have begun to speak of glacial lake hazard as a climate change impact. Luciano Lliuya even speculated that media attention related to his lawsuit pushed authorities to expedite construction on the early-warning system for Palcacocha that was finally completed in 2021. In the following chapter, I analyze how lake workers' engagement with the environment at the Palcacocha flood safety project expands the stakes of glacial politics.

9

Engineering in a Sentient Environment

It was a rainy morning in March 2017 when I first witnessed Eduardo Díaz speaking with the mountains. After traveling to the lake with Luciano Lliuya and a German journalist, I had spent the night at the workers' camp. Frosty temperatures and high altitude made for fitful sleep, and I spent much of the pitch-dark night listening to the workers snoring. Although the laborers normally awoke at dawn, the rain kept them in bed. Without equipment for working in bad weather, they could make no progress on building the road to Palcacocha. Over a filling breakfast of soup and oatmeal, Díaz explained that he would perform a *pago* by the lake, which he did every two weeks. I was keen to watch.

The rain began to slacken as we made our way up to Palcacocha. Díaz carried a black plastic bag and skipped ahead at his usual swift pace. Climbing atop the concrete dam, we saw the misty blue lake. In the distance, glaciers stuck out from behind the low-hanging clouds. The journalist and I followed Díaz over loose rocks by the waterline up Palcacocha's right edge. With his bag, he walked about a third of the way up the lake, over half a kilometer.

FIGURE 9.1. Díaz addresses the mountains. (Photo: Alexander Luna)

On the way, he checked a measuring stick in the water to determine how much the water level had increased overnight with the rainfall. Finally, Díaz stopped at a secluded spot by the water. We waited several paces behind. In front of Díaz towered the great mountain peaks. He set down his bag and extracted some coca leaves, candy, and a small bag of sugar. It was time to perform the *pago* ceremony. Díaz turned to the mountains and addressed them in his native language of Quechua (figure 9.1): "Apu Pucaranra, Apu Palcaraju, your tears have formed what they call Lake Palcacocha. Your people live in this lake, and I have brought them their sweets: candy, sugar, different kinds of sweets, your coca. You told us that you wanted coca, so I brought it to you."[1] Díaz then threw a handful of coca leaves into the water. They floated on the lake's surface. He tossed in a few pieces of hard candy and watched them sink. He poured a portion of sugar out of the plastic bag into the lake. Finally, he retrieved some pastries from the bag. "You told us you want pastries, so I brought them to you." Díaz threw in the pastries between the floating coca leaves.

"So don't scare us anymore, don't make your mountains crash. That's why I've brought food for both of you, for you and for your people that live here. If you scare us, you will also scare those living in Huaraz, in Unchus, in Llupa, in Nueva Florida, and you will scare the people living near those places. I'm bringing you what you asked for so that you won't scare us."

Díaz retrieved a half-liter plastic bottle filled with a clear liquid, taking the cap off as he looked down at the lake. It was homemade cane liquor.

"Also have your alcohol, you told me: 'Bring me the other thing; I'm thirsty.' I've brought that as well; here it is, drink up."

He poured a dash of liquor into the lake and put the top back on the bottle.

"Don't scare us anymore, from now on we won't say: 'Pucaranra mountain is crashing down.' We will be calm, and will say, 'They're behaving; we've given them something to eat.' This is all I say to you, Apu; the next time you ask, I will come back."

With that, the ceremony was over. Díaz put the plastic bottle back into his bag and walked back up the lake. "Now, Palcacocha will be calm," he exclaimed to us. "It won't scare us anymore."

As foreman for the Palcacocha project, Díaz oversaw engineering works embedded in a techno-scientific logic that rendered the lake as a dangerous flood hazard. In his work, he also engaged with the mountains and lake through a logic based on trust and reciprocity. As I explore in chapter 3, social and moral relations with the landscape are an intimate part of life for many people in the Andes. For the lake workers, different logics and knowledges were inextricably entangled in flood safety infrastructure. The workers implemented scientific engineering practices while accounting for the environment's sentience. Through an ethnographic study of workers' engagement with the socio-material environment, I reveal the hidden knowledges and ways of being at stake in glacial politics. Building on the work of anthropologists who have documented the ways modern political systems systematically exclude ecosystems as participants, I show how mountains and lakes emerged as hidden political actors.

Flood Safety Infrastructure: Design at Its Limits

It was mid-2011 when Fernando Vilca received a phone call. He was a middle-aged engineer from Huaraz working on an irrigation project in the highlands near Lima. Over the phone, a friend told him to apply for an open job implementing flood safety works at Palcacocha. A new risk assessment by the Glacier Authority had found that the lake had quadrupled in size since the last measurement in 2003, and the Peruvian president had declared a state of emergency. While Vilca was initially reluctant to work for the Ancash Regional Government, which had a reputation for instability and corruption, he agreed to send his résumé by email. "So I got the job," he later explained to me. "They called me and said to come sign the contract immediately."

In July 2011, Vilca began working at the lake. He was supposed to complete the project in five and a half months, but quickly ran into unforeseen difficulties. The initial plan was to install six plastic siphons to reduce the water level by 15 meters before beginning work on the new dam. Once installed, each siphon was almost 700 meters long and around 25 centimeters in diameter. The siphons came in sections 6 meters long, each weighing seventy kilograms. Vilca hired men from nearby villages to work on the project. With only a footpath to access Palcacocha at the time, it was painstaking work to carry the siphons to the lake, piece by piece on the shoulders of two laborers. The project mandated fifty days to transport around seven hundred siphon sections. It actually took three months. "After that," he told me, "they said to start draining the lake."

As the siphons began arriving at Palcacocha, Vilca ran into the next problem. He was supposed to lay the siphons over a path 400 meters long descending from the lake. "To do that, you need flat terrain, free of everything," Vilca explained. "But the whole area was full of immense boulders." As the national park authority would not allow him to use machinery, they had to use hand tools. "We had to remove all those rocks. We had to break apart the big rocks with brute force, using a wedge, because they didn't let us use explosives in the beginning. Those are things that weren't considered in the project plan." Making slow progress, Vilca and his workers cleared the pathway and began installing the siphons. Rather than five and a half months, as originally planned, it took over a year until the drainage system was running, in August 2012. "It was calculated really poorly," exclaimed Vilca.

As the work continued, Vilca explained, it was difficult to find reliable workers. Few people could stand the hard labor of carrying heavy equipment to Palcacocha and spending weeks at the isolated lake. "After a month and a half, the majority seemed to get tired; they got bored with the job and abandoned their post." Vilca sought out Díaz, who was known in the area for his extensive knowledge of mountain plants and expertise in healing people with homemade remedies. He had also worked on glacial lake infrastructure projects in the 1960s and '70s, giving him valuable experience with dams and drainage systems. Díaz later recruited his son-in-law Martín Amaru, a mountain guide used to working at high altitude. While the regional government saw a high turnover in public officials over the subsequent years, Vilca, Díaz, and Amaru formed the core of the Palcacocha project. The latter two spent most of their time living at the lake, with Vilca visiting several times each week.

Infrastructure often fails because project planners approach nature as an orderly domain separate from society (Edwards 2003, 195). Engineers deploy epistemological techniques such as standardized measurement procedures to understand complex environments. But in practice, engineering involves a tension between project design and pragmatic engagement with unpredictable environmental conditions. Engineers are often acutely aware of the fact that the environment is less stable than their designs suggest. At times, they must adapt to unexpected challenges not accounted for by designs and abstract standards (P. Harvey and Knox 2015). The Palcacocha project design reached its limits when Vilca encountered socio-material resistance, such as heavy boulders and unreliable workers, that prevented him from installing the siphons as planned. He and his workers developed pragmatic solutions that accounted for social and material difficulties.

Once they had set the siphons in place and opened them in August 2012, Vilca and his workers began lowering the water level at Palcacocha. The siphons worked with gravity: The tubes ended at a lower altitude than the lake level, so once they started running, they continuously pumped out water. Díaz and Amaru made sure the siphons functioned properly, performing daily maintenance. They fixed punctures from falling rocks. Sometimes an avalanche would hit the lake, causing waves that swept the lake ends of the siphons out of the water. Using a little rubber boat, they put the siphons back into place. As the project funds began to run out, they bought a two-way radio and antenna with their own money, establishing a communication system in a small hut they built above the lake. They monitored Palcacocha twenty-four hours a day, providing regular updates to officials in the city via radio.

"We've gotten used to it," Amaru explained to me one morning beside the radio as we glanced toward the morning mist above the lake. "We've gotten used to the altitude, the cold, the sun's heat, the wind, everything. As responsible people we do this work; we're here permanently to monitor the lake's ascent and descent, reporting via radio to the authorities." Given administrative difficulties at the regional government, the workers often had to wait months until receiving their wages. It was a tough job, but Amaru was proud of his labor. The lake workers implemented scientific engineering practices to keep flood safety infrastructure running at Palcacocha. To continue functioning, the infrastructure depended on a complex set of relations: Lake workers and engineers employing techno-scientific standards and pragmatic expertise, scientists who rendered the lake as a flood hazard based on predictive modeling, and voices in the Huaraz population calling for urgent remedial works, as well as political institutions that had to fund flood safety infrastructure.

How do sentient ecosystems fit into the equation?

The Sentient Environment in Infrastructural Practice

Project foreman Díaz learned about the landscape at an early age. "My grandparents told me that the mountains and water are living beings," he explained to me. "From then on I knew that they are alive, just like humans." When he worked on flood prevention projects at other glacial lakes in the 1960s and '70s, he encountered local landscapes in his dreams.

Elaborate *pago* ceremonies are not a common practice in the Cordillera Blanca region. In his younger years, Díaz traveled all over Peru to work on construction projects. He learned about *pagos* for the landscape in the southern Andes, where the practice is still widespread, as anthropologists have documented (de la Cadena 2015; Stensrud 2019a). Some Andean and Christian practices have become entangled: The Qoyllur Rit'i pilgrimage in Cusco is a Christian ritual that involves appeals to mountain beings (Allen 1997). Díaz is Catholic, and his engagements with the environment are entwined with Christian understandings. In one interview, he explained that God could prevent a flood from happening at Palcacocha, "and if God wants, [the lake] can break out—nothing can stop Him." Engaging with the mountains requires a strong belief, much like his belief in God: "Only if you truly believe in them will they come to you."

Díaz's *pago* involved a reciprocal engagement with the mountains at Palcacocha. By providing them with nourishment, he hoped to maintain their trust and prevent them from causing a flood. This reflects similar practices in the wider region. Farther south is Lake Titicaca, which is much larger and lies on the border between Peru and Bolivia. In an ethnography of people's engagement with the lake, Ben Orlove (2002) has explained how the lake is not only a body of water and vital source of life for people and the ecology, but has myriad cultural and spiritual meanings. It is a central figure in local cosmologies and mythology, playing an important role in community identity. People interact with the lake as a living being, engaging in a reciprocal relationship marked by respect and reverence. In these ways, Titicaca is deeply interwoven with the social fabric and people's spiritual life.

For many Andeans, the landscape has agency. They must actively engage with that agency to ensure that it acts positively toward humans and their endeavors. People and powerful nonhuman beings are involved in mutually beneficial relationships. Mountains offer life and vitality in exchange for ritual offerings and other signs of respect. Positive reciprocal relations between people and landscapes ensure productive harvests. People gain knowledge and understanding of these beings through a continual close engagement with the environment (Allen 1988). Historians have traced Andean understandings about a sentient environment to Inca and pre-Inca symbolic geographies. The continuity of Andean understandings about landscape agency is remarkable given extensive historical change—from colonization, through land reform, to recent processes of neoliberal development. People continue to renew and reconstitute their relations with the environment, acknowledging its agency through reciprocal offerings (P. Harvey 2001).

As the engineer in charge of the Palcacocha project, Vilca had an almost managerial engagement with the mountains. Unlike the foreman Díaz, who grew up in a Quechua-speaking rural family of farmers, Vilca came from an educated family in the city of Huaraz. He saw *pagos* as an important aspect of hazard management at Palcacocha and frequently reminded Díaz to perform the ceremonies. "It's hard to believe," Vilca explained to me on one of our many bumpy rides up the road to the lake, "but when Díaz wasn't performing them, the water level went up; and when he did the *pagos*, the level went down." While they were building the road to Palcacocha, Díaz got sick and stayed in his village for several weeks. During that time, there were two accidents that left workers injured. Vilca had not believed in these Andean practices, he told me, until he saw them produce real effects.

Before going to university, he worked as a mountain guide. At the time, he did not believe that mountains were living beings. During one climbing tour, an avalanche nearly left him dead. Later, he concluded that the avalanche was a violent response from a powerful mountain because he failed to show it respect. Vilca gave up on mountain climbing after the accident and began his studies as an environmental engineer. Working at Palcacocha, he sought Díaz's support to maintain positive relations with the landscape and placed great value on these practices from an academic and philosophical perspective. Turning back to me from the front seat as we slowly rode up to the lake, he exclaimed, "We have to preserve this Andean science!" For Vilca, technical expertise, an understanding of the social relations at stake, and knowledge of the environment's sentience were all crucial to his labor at Palcacocha.

Díaz began working at Palcacocha in 2011. Vilca and the authorities in Huaraz were worried about the lake's high water level. When Díaz first slept in the camp by the lake, as he and the workers were installing the drainage siphons, the mountains and lake approached him in his dreams. They looked and dressed like local people. "You see me, I live by your side," he explained to them. "Don't scare me!" They told Díaz not to be afraid. "They said to me, 'I want coca, cigarettes, beer, and alcohol. If you bring all that, I will be fine; I won't scare you.'" Díaz began performing regular *pagos*, fulfilling the beings' requests, and the water level receded.

For Díaz, techno-scientific hazard management and a relational engagement with the landscape went hand in hand. "The siphons stop the water from growing," he explained in an interview. We sat in a dark hut by Palcacocha, next to a small fire that kept us warm. "They make the water level go down, and that makes me happy. The siphons have their own will; they work as if they were people. The mountains help as well; they are also alive. The lakes and the mountains—they are all alive if we have faith. They are listening to us now as we speak." His work would be successful as long as he showed the landscape sufficient respect. "I'm happy. The lake has shown me no resentment; it's cheerful. I always provide its *pago*, and my siphons are working day and night."

The lake appeared most clearly in Díaz's dreams. "The lake is five people. A woman, the mother, and a man, her husband; and their children—a plump woman and a thin one, and a guy who's in the middle. They come and talk to me; ask me how I am." Díaz felt a close connection to them. "They talk to me with great trust, like my family members. They talk to me like we're sitting here with you, and they tell me not to worry." Trust was key in Díaz's relation with the beings: "They're very fond of me." Yet their power was also clear to Díaz; hence his appeals that the mountains stop scaring him and prevent a flood disaster.

Many anthropologists have written about encounters between universalizing scientific knowledge and particularizing relational knowledge, often focusing on relations of power. In her study of people's engagements with the environment on the border between Canada and Alaska, Cruikshank (2005) has described how Aboriginal narratives account for glaciers as sentient beings. She contrasts local approaches that emphasize the landscape's agency and value mutual relations with the sentient environment with a scientific colonial approach that objectifies nature in an effort to control the colonized environment. In Díaz's daily labor, technical and relational modes of engaging Palcacocha became entwined. These different domains are not necessarily mutually exclusive; Indigenous knowledges are not always opposed to modern scientific knowledge (P. Harvey and Knox 2015). When farmers struggled with an unruly environment in the construction of irrigation infrastructure in the Cusco region of southern Peru, they offered both technical and relational explanations: While they accepted engineers' concerns that the environment's biophysical condition was the problem, they also argued that work had stalled as people had neglected to demonstrate sufficient respect for sentient beings in the environment (Boelens 2014).

Cruikshank (2005) has argued that stories about sentient glaciers can broaden public debates about environmental change. In revealing how the environment can be sentient and agentive, the stories enact an understanding of the world in which nature and culture are inherently entangled. Such perspectives can provide an alternative to dominant narratives of nature as an object to be managed by humans. Díaz came to know the lake and surrounding mountains by taking scientific measurements and by engaging the landscape in *pago* ceremonies and his dreams. According to its design, the Palcacocha project was embedded in scientific and engineering standards. Implementing this infrastructure in practice involved an engagement with the socio-material environment that evoked technical standards alongside pragmatic empirical knowledge.

Mountains Play Politics

What kept Vilca, Díaz, and Amaru working at Palcacocha, despite the difficult conditions? All three felt a strong sense of responsibility. Speaking to Amaru in the radio shack one sunny morning above Palcacocha, I asked him why he was still there, even though the regional government rarely paid him. "I'm here because I love my work," he explained. "I'm here even if they don't pay me, because work gives us dignity, and out of love for my people, for the city of Huaraz; because this is what any Peruvian should do."

The engineer Vilca quit his job for several months in 2015, but ultimately returned when regional officials promised to pay him back wages. They never paid the full amount, and the situation returned to normal: "Even if they aren't paying me, here we are, trying to make this work." Vilca felt dedicated. "Time goes by, years go by, but I can't abandon this situation. So we have to continue; not because we want to, but because the sociopolitical circumstances obligate us to do this work." He was glad that he could count on Díaz and Amaru, both of whom stuck around through all hardship. "They've taken this labor to be their own," he explained. "They see this as a service to society. The three of us have identified ourselves that way; that's why we endure it all even if they don't pay us." If they left, who would take their place? "There aren't any other people who could do this."

And the foreman Díaz? Despite being in his eighties, he felt a stubborn devotion. "I live here for the people of Huaraz; in the cold, with the frost—

I've gotten used to it. I endure being here because of what has been revealed to me, and so I pay and pay the *pago* to the lake." According to Díaz, his engagement with the landscape was crucial for preventing a disaster. "When I provide the *pago*, the mountain doesn't come down. But when I leave for my break after fifteen days, the mountain can fall down and scare us," he explained. "As long as I'm here, it's calm; the mountain doesn't collapse, and it doesn't scare us." While his relation of trust with the mountains was strong, there could be trouble with other people. "They [the mountains] said to me: 'If you leave me and another person comes, that person might die. We'll get rid of them.'" Díaz was worried: "They might cause a flood."

What role do sentient landscapes play in political engagements over Palcacocha? Can they participate in glacial politics, even if others refuse to recognize their existence? Those in political power often seek to set epistemological and ontological limits on which issues are relevant topics in political discussion (de la Cadena 2015, 276). In an interview, I asked an urban official involved in glacial hazard management about the claim that an avalanche had occurred because the lake workers failed to perform a *pago*. As an engineer, his answer reflected a scientific understanding of the environment. "People and populations have their beliefs," he explained. "Avalanches happen due to the mountain's geodynamics; we'll always have them." He said he had nothing against traditional practices or beliefs, but his perspective was realistic and rational: "Nature has its own dynamics, with or without a *pago*." From the urban official's scientific perspective, mountain beings were nothing more than traditional belief. Yet, for the foreman Díaz, they had a tangible presence. Different ways of knowing the Andean environment shape the stakes of glacial politics. But those participating in political disputes disagree about which knowledges are most legitimate for political decision making. The stakes of glacial politics depend on whom you ask. How can the politics of glacial retreat—and environmental politics more broadly—account for different ways of knowing and being?

Infrastructures are sites of engagement between people, materials, and nonhuman beings on one hand and the environment on the other. They shape what people perceive as social and natural (Jensen and Morita 2017). Infrastructural practices are often ambiguous, enmeshing different ways of knowing and being. Vilca as the head engineer did not engage directly with the mountains at Palcacocha; he left that to Díaz and Amaru. Nevertheless, he described the mountains as important actors in his work at the lake and reminded Díaz to perform regular *pago* ceremonies. Through these engagements, the sentient landscape came to figure among the stakes of glacial politics for Vilca and his workers.

In other political disputes in Peru, mountains have emerged as ambiguous yet contested actors. During social conflicts over mining projects, antimining activists have cited sentient environments to justify their concerns. In a controversial statement, President Alan García vociferously denounced premodern beliefs in mountain beings as slowing down national economic development. Despite García's refusal to accept them as real, mountains affected Peruvian politics when their presence forced the president to slow down mining expansion (de la Cadena 2015, 168).

This brings me back to an example mentioned in chapter 3. Fabiana Li (2013) has examined a conflict between farmers and a multinational mining company at Mount Quilish in the northern Peruvian Andes. Protesters described the mountain as a water source, using scientific terminology, and as an *apu*, a sacred mountain. While the latter characterization arose out of rural people's understanding that the mountain was sentient, describing it in this way was also a strategic move. Referring to the mountain as *apu* lent it a romantic appeal for urban environmentalists and journalists and spoke to contemporary imaginations of Andean indigeneity. In fact, the Quechua term *apu* was not common in that part of the country; activists had borrowed it from southern Peru. Mount Quilish came to matter in different forms through encounters between different knowledges. In asserting that both scientific and spiritual understandings were valid, activists sought to bring spiritual issues into the realm of the political (Li 2013).

While the politics of glacial retreat revolving around Palcacocha was less overtly conflictual, it also involved an encounter of knowledges and ways of being. In political discussions and infrastructural works, different sociomaterial engagements have emerged that point to a broad set of issues and relations at stake in glacial politics, and that come to bear on ethical claims about neighborly relations in times of climate change. In encounters with journalists and academic researchers—including myself—Vilca and Díaz described the mountains surrounding Palcacocha as *apus*, even though this term is not common in the regional dialect of Quechua. Díaz even used the term as a sign of respect to address the mountains in *pago* ceremonies. They may have picked it up on travels to other parts of Peru; perhaps they heard it in news reports about mining conflicts. Either way, the term is useful: It highlights the sentient environment's significance for political engagements about glacial retreat and climate change.

According to de la Cadena (2010), who builds on Latour (1993), modern political theory assumes a separation between nature and culture. Science mediates representations of nature through its knowledge practices. De la Cadena argues that sentient mountains can have no place in modern politics, as they constitute an alternative understanding, or ontological reality, of the environment. Modern politics can account for them only as cultural belief but cannot acknowledge them as real. De la Cadena argues that we should expand our understanding of politics to overcome the nature-culture dichotomy and allow for landscapes as political participants (de la Cadena 2010). Even when politicians ignore the role of sentient mountains, modern politics can involve ontological disagreement, as when mountains have come to participate in mining conflicts (de la Cadena 2015, 283).

How might politics come to include ecosystems as actors? According to de la Cadena (2010), the entry of landscapes into modern politics questions the latter's very foundations by disavowing the separation between nature and culture. Disrupting politics as usual, it can allow for pluralization. De la Cadena argues that a redefinition of politics can acknowledge the existence of multiple worlds or socio-natural formations.[2] A "new pluriversal political configuration"—a cosmopolitics—permits legitimate political disputes between different worlds (de la Cadena 2010, 361). De la Cadena calls for a politics that recognizes ontological disagreement across multiple worlds on the basis of mutual respect (de la Cadena 2015, 285).

At Palcacocha, the mountains and lake have inadvertently become part of political processes, even if not everyone involved in those processes recognizes them as actors. In the Palcacocha project, different ways of knowing and engaging with the Andean environment came to bear on political disputes. When Díaz made appeals to the mountains in his work at Palcacocha, he drew them into the stakes of glacial politics. While state officials in Peru and elsewhere have denied that sentient landscapes are relevant in political discussions, ethnographic study at sites such as Palcacocha reveals that nonhuman entities can nevertheless play a role, even if their role is mediated by human actors. Given this proposition, "modern politics," if conceived as relying exclusively on scientific forms of knowledge, no longer appears as a coherent approach. In political practice, different knowledges can emerge hand in hand to form parts of political claims about social relations and the environment. While their existence is ever ambiguous and often disputed, sentient landscapes play a significant role in Peruvian glacial politics.

How might this discussion about Lake Palcacocha inform broader conversations about climate change? As with Peruvian glacial politics, climate politics builds on scientific and other knowledges about environmental change at local and global scales. These knowledges shape the scope of political claims about how societies, governments, and corporations should respond to climate change. In his ethnography of environmental politics in Hong Kong, Timothy Choy (2011) argues that political claims gain strength when they address both specific and generalizable issues. In Hong Kong, environmental activists formulated their concerns about endangered species in local and global terms, relating situated worries to environmental problems at a larger scale. In his legal proceedings against RWE, Luciano Lliuya attempted a similar move in addressing both specific concerns about glacial retreat in Peru and global worries about climate change. As in political deliberations over Lake Palcacocha, his claim brought together diverse knowledge practices: from scientific knowledge about climate change and flood risk in the Andes, through legal knowledge about causal accountability, to Andean ways of knowing that account for environmental agency. These knowledges shaped the potential scope of social and political discussions. The German judicial system does not recognize mountains as relevant legal actors, as I discuss in chapter 3. Nevertheless, for Luciano Lliuya the claim was an attempt to expand the stakes of climate change politics: to account not only for scientific conceptions of global warming but also for the landscape in its own right.

Luciano Lliuya engaged RWE and the Andean mountains as different kinds of neighbors. While his relationship with RWE emerged out of scientific evidence and the legal norms of German neighborhood law, his connection to the mountains was based on an intimate lifelong engagement with the Andean landscape. Both RWE and the mountains are powerful actors, albeit in very different ways. From RWE, he demanded financial support based on its moral obligations as a contributor to climate change. His relationship with the mountains was not legally codified; it emerged out of a deeply felt sense of reciprocal moral obligation. If he shows the mountains respect, he hopes they will treat him well. Seeing the mountains suffering, he felt a moral obligation to take action on their behalf. Like other Andean villagers, Luciano Lliuya has expressed uncertainty over what the mountains are and how glacial retreat might affect them. Nevertheless, he hoped to bring them into conversation as an additional actor at stake in global climate politics.

FIGURE 9.2. The engineer Vilca at Lake Palcacocha. (Photo: Alexander Luna)

* * *

I RETURNED TO Lake Palcacocha in May 2022 after almost four years away. I had been busy working on my PhD, and later the COVID-19 pandemic restricted global travel. The German court had begun planning a visit to Peru in 2019 to gather evidence at Palcacocha, but it was repeatedly postponed. The pandemic hit Peru hard. Its underfunded public health system was quickly overloaded. There was a shortage of oxygen: People had to buy oxygen canisters for their sick family members on the black market for thousands of US dollars. The region of Ancash was especially badly affected. When the public hospital in Huaraz ran out of beds, authorities converted the city's stadium into a makeshift clinic.

The pandemic delayed both the lawsuit and safety works at Palcacocha. In 2021, authorities finally completed an early-warning system. They installed sensors at the lake that should detect an outburst flood. Sirens have been mounted throughout the city and surrounding villages to warn the population and initiate an evacuation. Authorities organized workshops with communities to raise awareness. As of 2025, little progress had been made on the project to build a new dam and drainage system at Palcacocha, though authorities were beginning to examine the prospects for a project addressing both flood hazard and water security.

The foreman Díaz was let go in 2018. His son-in-law Amaru continued working at the lake until 2022 and took on the duty of performing *pagos*. Several large avalanches have hit the lake in recent years, causing significant public worry. Luckily, none of the avalanches have been big enough to cause a flood downstream.

The Palcacocha engineer Vilca was let go by the regional government in July 2020, allegedly for not supporting the governor's political party. They owed him over a year's worth of pay, and he went to the government office almost every day wearing a respirator mask to demand his money. News reports later said that he likely contracted COVID during those visits. He was interned at the hospital and passed away in April 2021. He left behind his wife and son. He was a conscientious engineer and proud Andean whom I was lucky enough to count as an interlocutor and friend. He made great personal sacrifices to keep the people of Huaraz safe. "I studied at a public university," he once told me, "and that means I have the duty to give back what I can to society." I dedicate this book to Vilca: He lived and died for Lake Palcacocha (figure 9.2).

Interlude 4

Unexpected Stardom

"It went perfectly. A super decision. Now it begins!"

On a cold Andean morning in late November 2017, I received a WhatsApp message from Luciano Lliuya's lawyer. The judges had issued the ruling moving the case into the evidentiary stage. She texted me from the courthouse in Hamm after a brief hearing that formally confirmed what the judges had indicated several weeks earlier when Luciano Lliuya was in the courtroom: They found that the lawsuit was admissible and wanted to examine evidence to see whether evidence of a causal link between RWE's activities in Europe and climate risk to Luciano Lliuya's property in South America could meet the appropriate legal standard.

If this was the beginning, where would the case end? The lawyer's words gave me cause for reflection as I stood inside my dark little room in an adobe house, my home during much of my fieldwork in the Peruvian Andes. I had eaten a quick breakfast with my host family before they set off to the fields for the day's work. Their children had already left the village in a rickety public bus to attend school in Huaraz, in the valley below. Under the icy morning sun, I drove off on my motorcycle to meet Luciano Lliuya. He had spent the night higher up in the mountains. He was seeing to his cow that roamed freely, grazing in a grassy valley, and was about to give birth.

By that point, the lawsuit had already achieved much more success and recognition than we had ever imagined it could. In early discussions with Luciano Lliuya, the lawyers told him that the chance of legal victory was approximately 10 percent. The court could at the outset have rejected the claim, which appeared to many as outlandish. As the first case of its kind, even basic evidentiary standards were in question: How could a court measure causal impacts of global climate change? Yet the judges in Hamm decided to take the case seriously in all its complexity. They acknowledged that it could be possible under German law to hold greenhouse gas emitters accountable for their contribution to climate change impacts.

I met Luciano Lliuya at the end of a bumpy dirt road at the mouth of a wide valley leading up toward the glaciated mountain peaks. He was happy to hear the news, responding in his usual reserved manner with a slight smile. We speculated how the case might continue. Luciano Lliuya had already received significant media attention. He reflected that in terms of *la causa* (the cause) we had achieved significant success by bringing people's attention around the world to climate justice and glacial retreat in the Andes. In the long run, Luciano Lliuya hoped this would make a small contribution toward stopping global warming and saving the mountains that surrounded us from losing their white caps.

On the motorcycle, we drove down the road looking for cell phone reception. Finding a place that gave us a view of Huaraz in the glorious morning sunshine, we called Luciano Lliuya's lawyer to hear more details. She explained that the court hearing that day had been short and straightforward: RWE's lawyers had presented new written arguments several days earlier seeking to undermine the lawsuit's merits, but they had not convinced the judges. The case would likely go on for several years. The lawyer was thrilled.

Standing by the road with the mountains behind him, Luciano Lliuya wondered what this turn of events would mean for his immediate future. He had been celebrated on the international stage as a hero of climate justice. During press interviews, journalists from around the world often expressed open admiration for him and his claim. Luciano Lliuya is a shy person who never intended to become famous; he took on the role of a climate change celebrity with initial reluctance. In his own view, he was not—or should not be—at the initiative's center. He was only acting on behalf of the mountains.

As we conversed beneath the mountains, I received a call from a producer at a German TV station. They conducted a brief video call with Luciano Lliuya that was later broadcast to households around Germany. Luciano Lliuya explained that he was happy with the result. No matter how the case ended, the fact that his cause had received so much attention already felt like a massive victory. After the interview, I drove Luciano Lliuya back up to the valley—he had to check on his cow again.

Conclusion

Changing the Legal Climate

Luciano Lliuya's neighbors were surprised when a crowd of fifty people descended on their community in May 2022. The group consisted of representatives from some of the most important national and global media outlets, including *The Guardian* and the *Washington Post.* Camera crews from four documentary film teams followed Luciano Lliuya and his entourage around. Luciano Lliuya, his lawyer, and I led a tour through the city, showing journalists the flood danger zone, evacuation routes, and early-warning sirens. We ended the tour on the roof of Luciano Lliuya's house, where he gave interviews until dusk fell on Huaraz (figure C.1). The media hurricane arrived ahead of an unprecedented legal milestone: The German court was visiting Huaraz to take evidence in Luciano Lliuya's claim against RWE.

In the past, Luciano Lliuya had faced negative rumors in his own community. Some of his neighbors questioned the outlandish lawsuit and speculated that he must be up to no good. The arrival of the Germans made the claim look more legitimate to local eyes. It was no longer something happening far off in a foreign country; now German judges were in Huaraz to inspect the melting glaciers. Luciano Lliuya's legal team met with local community groups that shared their concerns: If their plight came to global attention, the locals argued, it should benefit the whole community, not only Luciano Lliuya. His lawyer explained that the case was about setting a legal precedent that aimed to help all communities in the world affected by climate change. By the end of the discussion, community leaders were satisfied, and they welcomed the judges when they arrived at Lake Palcacocha several days later.

The court visit involved a large official delegation: two judges, five court-appointed experts to take evidence, three lawyers for Luciano Lliuya, four lawyers representing RWE, four experts advising Luciano Lliuya's team

FIGURE C.1. Luciano Lliuya speaks to the press at his house in Huaraz, May 2022. (Photo: Alexander Luna)

(including myself), three scientists with RWE, a medical doctor, and an interpreter. The lawsuit made climate change a moral issue, asserting a neighborly relation between Luciano Lliuya and RWE. The visit brought RWE's representatives into Luciano Lliuya's neighborhood and forced them to confront the reality of glacial retreat in the Andes. While climate change arises from encounters between people, ecosystems, and environments at an overwhelming planetary scale, the lawsuit made those engagements graspable. It configured climate change in terms of specific relationships between emitters, such as RWE, and those who face potential harm, like Luciano Lliuya. Encapsulated by its lawyers, RWE arrived in Huaraz as another contentious neighbor.

Luciano Lliuya rose to unexpected international stardom in an unprecedented legal claim. For the German climate activists who organized the lawsuit, he conveniently fit the role of a subaltern subject in the Global South who faces the worst impacts of climate change. Yet Luciano Lliuya did not present himself as a passive victim. He led a suit that offered legal, moral, political, and cosmological perspectives on climate change and social justice. Luciano Lliuya and his lawyers were surprised at the success the claim achieved, yet structural sociopolitical change addressing climate change was still far away. Crucially, the claim appeared to produce little immediate benefit to help Luciano Lliuya and his compatriots face the impacts of climate change.

I began this book with a question that Judge Rolf Meyer posed in a 2017 hearing: "Is [it] just" to leave people in poorer parts of the world on their own to face climate change, "even when we are causing the problem over here?" The matter at hand was strictly legal. The judges had to decide which norms of justice should regulate the relationship between RWE in Germany and Luciano Lliuya in Peru. Yet the judge's question was both moral and political: He broadened the scope of discussion to the relationship between "the places in the world where money is scarce" and "we" who "are causing the problem over here." The judge made explicit what many observers had already speculated: The lawsuit concerned not merely a connection between localized human and corporate persons but the very fabric of global life in times of accelerated environmental and social change. At stake was the relationship between the "Global North" and "Global South." The claim reconfigured climate change by asserting a neighborly relation between RWE and Luciano Lliuya, and it called for collectively reordering relations between those who had made the largest contributions to greenhouse gas emissions and those who face the worst effects of climate change.

This book approaches the lawsuit as an opportunity to examine how climate justice claims enact morally charged relations that draw together people, corporations, and other beings. My starting point is the claim's fundamental ambiguity in appealing to both individual and collective moral responsibility. Luciano Lliuya, in collaboration with his lawyers and supporters, drew on disparate knowledges, perspectives, and moral norms to raise fundamental questions about how people should live and engage with one another on a warming planet. Each chapter highlights different dimensions of the social relations at stake in concerns about climate change in the Peruvian Andes, both within and beyond the legal process. I offered a set of perspectives gleaned from shadowing glacial lake workers in the high Andes, acting as Luciano Lliuya's interpreter in countless conversations with lawyers and journalists, and spending long hours poring over legal documents. These insights arise from my participation and commitment to the cause of climate justice. This book is the story of how a Peruvian farmer and German legal activists entered the conceptual sphere of climate justice and tried to make sense of it all.

Climate Change on Trial

What new perspectives does this book offer on creating responsible social relations in times of global warming? I began with the story of Luciano Lliuya facing RWE in court in order to show how the claim configured climate change in terms of neighborly relations. Taking analytical inspiration from this conception, I suggest that a neighborly approach allows for a fruitful ethnographic examination of climate change that focuses on how people make and contest moral relations. Building on academic discussions about climate litigation in law and sociolegal studies that unpack the social dynamics of legal mobilization, this ethnographic study uncovers the practical challenges of an emblematic climate justice lawsuit. In part 1, I explore who might be involved in moral engagements around climate change: While the legal framework permitted a claim between Luciano Lliuya and RWE as legal persons, reinforcing historically situated notions of human and corporate personhood, Luciano Lliuya raised the possibility that Andean mountains might also have a stake in the case. Other research has documented how modern legal systems exclude sentient legal systems from judicial and political deliberations; I show how they can play a role even without formal recognition. Luciano Lliuya engaged the mountains as ambiguous moral beings who are suffering due to climate change and glacial retreat. One of his primary motives in the lawsuit was to take a stand on their behalf.

Part 2 dives deeper into the judicial dispute before the German courts, examining how lawyers used scientific evidence to establish and contest a neighborly relation between Luciano Lliuya and RWE. A key issue at stake in the lawsuit was causality. I trace the legal practices that linked greenhouse gas emissions from RWE's factories to global warming, global warming to glacial retreat in the Andes, and glacial retreat to the risk of flooding to Luciano Lliuya's house. Scholars in sociolegal studies and STS have studied the use of science as evidence in court. I show how this plays out when cutting-edge science meets novel legal argumentation, examining how lawyers translate scientific facts about climate change into legal arguments. Claims about neighborly relations in times of global warming arise out of scientific knowledge that offers an increasingly detailed understanding of climatic processes. I call for more academic interrogation of causality claims that enact morally charged relations across geographic and conceptual bounds.

Finally, after a journey through courtrooms and legal texts, I return to where it all began: the Peruvian Andes. The lawsuit drew attention to broad questions of global politics, potentially overshadowing the socio-material relations at stake in Luciano Lliuya's home region. Focusing on the politics of glacial retreat, I show how concerns about glacial lake flood risk and water scarcity have emerged in relation to scientific and political engagements with the environment. When authorities implemented engineering projects to address flood risk, lake workers sought to appease powerful mountains and lakes as they executed an infrastructural project grounded in scientific knowledge. Ecosystems, I argue, are hidden actors in the politics of glacial retreat.

Where does all this leave us? Taking an ethnographic approach, this book traces the enactment of moral bonds between diverse entities with a prospective stake in the politics of climate change. Luciano Lliuya and his supporters drew on scientific climate knowledge and archaic legal theory to make the normative claim that polluting corporations and impacted humans are neighbors. We followed Luciano Lliuya, a reserved yet intensely principled protagonist standing up for climate justice, as he traveled between Andean mountains and German courthouses. We met RWE, materially manifested in its towering buildings and well-dressed lawyers, fighting with endless resources to unmake a sociolegal relation with a soft-spoken Andean farmer. And as they lurked behind the curtains, we caught glimpses of powerful Andean mountains and lakes—always difficult to comprehend analytically, yet intensely significant for those who directly engage them, and perhaps also important for those of us who do not. It makes for a fascinating narrative, but how might this story be relevant for other scholars and nonacademics looking to make sense of climate change? I end with some reflections on litigation, climate politics, and anthropology.

Neighborly Relations on a Burning Planet

In the lawsuit against RWE, Luciano Lliuya and his supporters invoked the idea of neighborliness as a normative claim about who should take responsibility for climate change. I use this claim as an analytical starting point to examine the ambiguous moral stakes of climate change. Starting with the normative concept, I offer some reflections on the broader relevance and repercussions.

Climate activists use the concept of neighborliness to highlight the responsibility of individual companies to individual plaintiffs, as well as that of all polluters to all those affected. This ambiguity can be productive in that it helps people conceptualize climate change in terms of a familiar moral framework, yet it also involves inherent contradictions. In the November 2017 court hearing, one of RWE's lawyers reflected on the implications of expanding the judicial concept of liability to cover climate change. He mused that "all of us in this courtroom" could be held liable. If the claim set a precedent, "there would be a wave of lawsuits by everyone against everyone!" This would amount to a totalization of individualized responsibility. The lawsuit takes place within a broader conversation about how responsibility for addressing climate change should be balanced between individual and collective scales. On one hand, there are efforts toward the responsibilization of individuals. Around the world, citizens are urged to fly less and avoid eating meat. This reflects ideas of responsible consumerism in an individualistic neoliberal framework. But while individual efforts are commendable, especially as a means of raising public awareness, they are unlikely to stop climate change on their own. On the other hand, many argue that climate change requires collective action by governments and corporations.

By invoking the idea of neighborliness, the lawsuit addressed the normative conundrum of individual and collective responsibility. Focusing on relations between individual actors risks drawing attention away from the broader power relations and structural dynamics of neoliberal capitalism that shape our contemporary era. Looking forward, climate change certainly requires global political action. It will not be resolved only through claims by impacted people against polluting companies. It requires solutions beyond the scope of individual relations, solutions that address broader social, political, and economic forces. The lawsuit is an attempt to enact change on a collective scale through a strategically targeted claim about individual responsibility.

While the legal framework places tight constraints on arguments and relevant knowledges, and while the lawsuit addressed only one specific climate change impact in relation to one emitter, litigation is a strategic tool to challenge climate politics as usual. By framing climate change in terms of neighborly relations, the lawsuit opened up the possibility of new political conceptions of responsibility in relation to climate change. The claim between Luciano Lliuya and RWE reproduced an individualistic legal framing that characterizes responsibility in terms of a relationship between two neighbors; yet for all those involved, the case concerned much more.

Nicole Rogers argues that legal structures were developed by ruling classes to defend their interests, yet disempowered groups can mobilize those structures to offer a radical critique of society. People can employ legal tools to challenge dominant social meanings and political structures (Rogers 2013). As an activist strategy, litigation harnesses the apparatus of the state to achieve political change, often combining old laws with new insights.

Activists use the tools of the establishment to challenge the establishment (Peel and Osofsky 2015). Strategic litigation has a significant deconstructive potential that can open up new political possibilities (Rogers 2013).

While the case framed climate change as a global process with local impacts, it also expanded that conception by drawing a moral bond between an emitter and an impacted person. Establishing this connection within a scientific and legal framework produced a power relation between Luciano Lliuya and RWE that, though unequal, introduced a new dynamic into social and political discussions about climate change. Climate change disrupts law in that it challenges dominant understandings of fundamental legal concepts such as liability (Fisher et al. 2017). Climate litigation, in turn, has the potential to disrupt climate politics by reframing discussions about responsibility. A focus on neighborly relations draws attention to the way litigation challenges dominant power structures.

As a multiscalar process in terms of causes and impacts, climate change is a significant regulatory challenge for policy makers (Osofsky 2007a, 234). Conventional climate politics involves a hierarchical approach to scale. From this perspective, climate change is a global process with wide-ranging local impacts. The political impetus for addressing climate change is focused on the global community of nation-states, with policies being implemented at lower levels of government. Political projects often build on particular scalar configurations that shape policy approaches (MacKinnon 2011).

In climate politics, scalar configurations arise from scientific conceptions of climate change. Drawing on an ethnographic study of climate modeling, Anna Tsing (2005) argues that climate models generate the notion of "globality." They aggregate weather and environmental data from around the world to produce global and regional models of current developments and future scenarios for climate change. These models convey a unified image of the globe, constructing the necessity to save the global sphere through unified political action at an international level (Tsing 2005, 103). Following this scientific approach, the global policy regime on climate change constructs the planet and atmosphere as regulatory objects (Whitington 2016, 9). For over three decades, the central international forum for climate change policy making has been the United Nations Framework Convention on Climate Change (UNFCCC) with its regular international conferences. At these summits, representatives of the world's governments seek consensus on how to stop global warming and deal with impacts that are already occurring. Numerous nonstate actors are involved in the process as nonvoting observers: NGOs such as Germanwatch participate in UN summits, attempting to lobby governments for more ambitious action. Some Indigenous people's organizations attend these meetings, as do energy industry representatives. Nevertheless, nation-state representatives dominate the relations of power, and unsurprisingly, large actors such as the United States, the European Union, and China wield substantial influence. In this political framework, government officials negotiate how responsibility should be distributed and who should pay how much to alleviate the impacts of climate change. Representatives of poorer countries have frequently demanded that large historic emitters foot a larger bill, as I discussed earlier. Unsurprisingly, the latter countries have been reluctant to comply.

While discussions at the UN have often concerned the role of nation-states, legal claims such as Luciano Lliuya's revolve around the moral responsibility of private corporations. In the contemporary capitalist political economy, business stands conceptually apart from the sphere of public politics: Policy makers establish a legal framework within which corporations can act of their own accord in the pursuit of profit. Corporations are not politically accountable to any population; instead, they answer to their shareholders. Under neoliberal global capitalism, corporations have become increasingly powerful actors, while accompanying discourses promoting individual self-advancement have shifted responsibility for social well-being from states to individuals (D. Harvey 2005). Corporations are legally configured as individual persons, yet their actions have much wider-ranging consequences than those of most human persons.

Thomas Eriksen (2016) has argued that people impacted by climate change face a scalar disconnect when they demand to know who is to blame. Following dominant discourses, they may assign responsibility for climate change to the generic global scale. This absolves any individuals, institutions, or organizations from responsibility (Eriksen 2016, 141). Climate litigation against polluting corporations offers a different approach. While climate change was previously characterized as a diffuse problem with numerous unidentifiable sources, scientific advances make it possible to identify specific actions and choices by individual actors that cause measurable damage. This approach turns climate change from a broad political question into an issue of individual concern and liability (Ganguly et al. 2018). Luciano Lliuya's lawsuit, building on climate change attribution science, invoked a moral bond between Luciano Lliuya as a person affected by climate change and RWE as a partial contributor. Drawing this new connection, and legitimizing it in the legal sphere, has profound implications for global climate politics: It recasts the relationships at stake. Climate change is no longer merely a confrontation between locals impacted and a global process, to be resolved at an international level among nation-states. Climate politics is recast to include translocal neighborly disputes between corporate contributors to climate change and people impacted around the world. This opens up a new power dynamic at the heart of climate politics—between affected people and large emitting corporations.

While the concept of neighborliness is understandable to many people around the world, that does not make it universally applicable. For many, it may serve as a helpful approach for thinking about how climate change connects us all around the world. Yet others will be reluctant to accept the abstract notion that they have a moral relationship with a corporation headquartered on another continent. Luciano Lliuya's lawsuit promoted the idea that corporations are moral actors. Yet many have objected to the capitalist legal fiction that corporations are persons with inherent rights and obligations. Refusing to accept this legal fiction undermines the idea that corporations can and should be good neighbors, or even that they can act of their own accord. Others will reject the idea that Andean mountains exist as moral actors. The normative argument about neighborliness posits that humans, corporations, and mountains are all legitimate political actors.

Climate litigation allows a wide range of actors to interact with legal and political processes concerning climate change (Peel and Osofsky 2015). Through the lawsuit, Luciano Lliuya became a significant actor on the stage of international climate politics. While claimants in climate litigation cases are typically located in a particular place where they face climate change impacts, making a claim connects them to other people around the world who are also impacted by climate change. This leads to a modified political understanding of climate change, one in which nation-states are not the only primary actors (Osofsky 2005). Global warming can give rise to new imagined communities around climate change concerns (Jasanoff 2010). Through its neighborly approach, the lawsuit promoted an understanding of climate change that links localized greenhouse gas emissions directly to localized impacts. It simultaneously invoked normative conceptions of individual and collective responsibility.

Neighborliness as an Analytic

What about the analytical value of neighborliness? Anthropologists have a long-standing interest in how moral relationships are constructed and contested. The neighborliness approach addresses the theoretical and methodological challenge of studying moral relations across immense scales. Global forces such as climate change, capitalism, and COVID-19 have brought us all closer together. But how exactly? What kinds of moral responsibilities arise from these processes? Some have promoted universalist ethical frameworks such as cosmopolitanism, arguing that we all have the same rights and responsibilities (Beck 2006). Marxist critics point to the power relations and historical inequalities that shape global engagements (D. Harvey 2009). Others bring an ecological perspective to the table, calling for more attention to humans' entanglement with the Earth and its ecosystems (Latour 2021). The neighborliness analytic traces how these perspectives can in practice become entwined as people try to make sense of who should solve the world's problems. Ensuing moral claims may be ambiguous and even contradictory, highlighting the responsibility of divergent actors across geographic and conceptual scales. As a research framework, neighborliness unpacks these nuances to understand how people make sense of moral responsibility on an increasingly interconnected planet.

Activist claims often seem straightforward: They involve moral arguments about who is suffering and who is to blame. These stories offer simple explanations of complex social and material realities. The neighborliness analytic helps us untangle those underlying realities. It provides a grounded and relational approach to understanding claims about moral responsibility. It offers a way to conceptualize how humans, corporations, and ecosystems are bound together. Neighborliness is a familiar concept that resonates with people across social and geographic contexts, making it an accessible entry point for discussing complex global issues. The analytic offers a perspective for studying moral relations across scales: from the individual to the collective, from the local to the global. Climate change gives rise to new social links between people, corporations, and other actors over immense distances. Neighborliness helps theorize global interconnection in an era of climate change, tracing how moral relations are constructed. Studying climate change in terms of neighborly relations highlights the ambiguous and productive tensions inherent in moral claims about who should address the problem at different scales. This approach highlights ambiguities at various levels: in terms of the scale of moral responsibility (such as individual or collective), which actors are enmeshed in relations of responsibility (who is responsible to whom), and what responsibility means (how good neighbors should act).

The neighborliness analytic addresses some of the theoretical and methodological challenges that arise when analyzing a world-encompassing process from a social research perspective. This approach traces the moral relations that emerge from scientific conceptualizations of climatic processes and that people invoke to make sense of how global warming connects major polluters and those who face dramatic environmental transformations. It highlights the role of NGOs, governments, nonhuman beings, and other potential stakeholders caught up in concerns about climate change. A focus on neighborliness points to the uneven power relations in social, political, and legal discussions about climate change, and captures people's attempts to reshape those relations. Luciano Lliuya and RWE formally engaged each other as equals in the legal context—as legal persons with institutionally defined rights and responsibilities—yet the company undoubtedly had more resources at its disposal than the plaintiff and his activist backers. As Luciano Lliuya's experience in his village shows, neighborly tensions can revolve around whose voices count and how resources should be divided. If neighbors are those who can potentially cause harm to each other, we might say that all people, corporations, and state institutions in the world are potential neighbors. Nevertheless, neighbors become neighbors only through moral claims that assert a concrete relation between clearly defined beings or entities. My approach is to follow these claims as they emerge.

Going beyond this specific case, what is the broader value of the neighborly approach for anthropological and sociolegal analysis? And what are its limitations? A focus on neighborliness highlights the contested nature of social relations in times of increasing global interconnectivity. The bounds of mutual moral responsibility are negotiated through legal and political claims about who should address the major challenges of our time. The concept of neighborliness can be applied to a wide variety of relationships. The term highlights the moral stakes of social relations across geographic and conceptual scales. Not all neighborly relations are of the same sort. While Luciano Lliuya's engagements with RWE, mountains, and his village neighbors all involve moral concerns, they emerge out of very different social and material practices. None can be equated with the other, and each must be examined in its own context.

While a focus on neighborliness is helpful for understanding the moral dynamics of a global issue like climate change, an appreciation of the broader sociopolitical-economic context is also needed. An analysis of moral relations should not distract from the broader structural forces at play, such as capitalism, inequality, and racism. Rather, the neighborliness analytic can serve as a tool for examining how those structural forces operate and are challenged through social engagements and moral claims. Care must also be taken not to oversimplify: A neighborly analysis does not entail calling all relations neighborly but involves studying how moral relations are constructed across scales. The bounds of moral relations are ever shifting.

The neighborliness concept is a tool for understanding how people come to understand climate change as not merely an atmospheric process with countless moving parts but a moral issue with identifiable perpetrators. Future studies might examine how climate litigation against governments engages these public entities as different sorts of neighbors in arguing that they should take more ambitious action to tackle climate change. Rather than establishing a neighborly relation in the first place, as with the lawsuit between Luciano Lliuya and RWE, such claims seek to recast existing relational responsibilities between citizens and governments. Recent youth-led environmental protests, most prominently involving Greta Thunberg, concern ethical responsibilities between older and younger generations. While this book focuses on claims about responsibility across geographical and relational scales, future analyses could focus on the role of time scales. Time featured in the lawsuit against RWE, which concerned the company's historical responsibility, and plays a critical role in other cases, which argue that governments and corporations should change their future behavior to make the planet more livable for humans and ecosystems.

A focus on neighborly relations traces how people come to understand the broader issue of climate change in terms specific to relational engagements. Going forward, this approach could be applied to other environmental disputes. For example, water scarcity is an issue affecting increasing numbers of humans and ecosystems around the world. Rivers connect people and places across large distances, and water availability is shaped by industrial and agricultural processes as well as the whims of the environment. A neighborliness approach could be used to examine how such disputes play out through social engagements and moral claims between humans, social structures such as governments and corporations, and ecosystems. The neighborliness analytic could also be used to study other, socio-material processes that transcend scale, such as capitalist exchange, migration, and public health. The COVID-19 pandemic is a case in point: A novel virus—another actor of some sort?—connected nearly all humans on the planet and created shared experiences of illness, isolation, and mutual support. Governments, citizens, and corporations were all responsibilized to limit the pandemic's spread. A neighborly approach can trace how COVID became a matter of both individual and collective responsibility, as arguments raged in the political and public spheres about how responsibility should be distributed.

The neighborliness analytic provides a novel conceptual framework with which to examine moral relations among people, corporations, governments, and ecosystems connected through globalizing forces. It helps unravel the nuances of normative arguments, tracing how they emerge; who they involve across geographic, temporal, and conceptual scales; and how they are deployed in practice. It does not provide moral answers; rather, it can inform our normative commitments by helping us understand the composition and implications of moral arguments.

Expanding Perspectives Beyond the Courtroom

Climate litigation raises fundamental questions about how people should engage with one another and their environments and who should take responsibility for increasingly catastrophic disruptions. Courts have become sites of dispute about these issues, offering strategic platforms that give a voice to subaltern actors and allow them to participate in public discussions. Courts can rule on individual cases. Nevertheless, addressing these questions in the long term requires social negotiation over and political decision making about which values should be at the forefront in addressing climate change.

Going forward, how might social research contribute to these discussions? In their review of academic literature on climate litigation, Joana Setzer and Lisa Vanhala (2019) assert that social science research offers a productive approach for understanding the broader social significance of climate litigation. The unique contribution of this book is an ethnographic study of how climate change is brought to court, placing litigation in the context of social and political discussions about climate change at local, national, and global scales. I show how knowledge frameworks and ontological standpoints shape how people understand and engage with climate change. Litigation frames disputes in terms of legal and scientific technicalities, yet judicial claims are shaped by power dynamics that often leave litigators struggling to cover court costs while polluting companies possess seemingly infinite resources to counter legal claims.

What happens to the standpoints that the judicial framework excludes? When social justice activists take legal action, they subject themselves to a restrictive epistemological and ontological politics. In epistemological terms, I demonstrate throughout this book how legal procedure systematically leaves out nonscientific perspectives. Lacking scientific or legal qualifications, Luciano Lliuya made little formal contribution to the written arguments brought forward with his claim. He merely offered anecdotal evidence that held limited sway in legal proceedings. In ontological terms, the judicial framework did not recognize the Andean mountains as legal actors; yet Luciano Lliuya's relationship with mountain beings under threat from climate change was the principal factor motivating his participation in the claim. As I discussed in chapter 3, some legal systems have begun to expand their ontological horizons, granting legal rights to rivers and mountains. Alternative standpoints may achieve increased public and political recognition as a result of climate litigation, even when legal frameworks fail to acknowledge them.

As I argue in this book and others have demonstrated,[1] legal proceedings can inadvertently grant public legitimacy to subaltern ontological understandings. Social justice–oriented legal claim making involves more than judicial process. Campaign efforts and public discussions around lawsuits are crucial to activists' efforts. Many of the most significant impacts from Luciano Lliuya's claim arose far beyond the courtroom, including accounts in *Time* Magazine (Nugent 2018), the *New York Times* (Jarvis 2019), and the *Washington Post* (Kaplan 2022). While judicial procedure offered few opportunities for Luciano Lliuya to contribute his perspective, media reports foregrounded his experience of climate change in the Andes. His perspective was legally anecdotal, but existentially significant. The *Time* Magazine article begins with a dramatic snapshot: "Climbing a snowcapped mountain in the predawn light, Saúl Luciano Lliuya says he could sense something changing. All his life, pristine glaciers have nestled between the peaks surrounding his hometown in the Cordillera Blanca region of the Peruvian Andes, providing water, work and beauty. 'Now you can see it,' he says. 'They're disappearing'" (Nugent 2018).

This points to a multilayered epistemological politics at the heart of climate litigation: While the judicial standards of evidence are strict and exclusionary, situated experiential perspectives appear alongside scientific facts in media articles. Luciano Lliuya's claim generated wider public sympathy by placing a human story at the center of climate change discussions. In this sense, Luciano Lliuya made a significant epistemological contribution to the legal case.

Luciano Lliuya asserted that he should not be at the lawsuit's center, but rather the mountains themselves. The mountains had no legal standing in the courtroom. Within the ontological framework of the German judicial system, they did not exist as actors. Luciano Lliuya brought Andean Earth beings into public discussions on some occasions, yet it remains to be seen whether this inclusion will have a wider impact. It could expand the scope of climate politics by bringing new actors into play.

Engaged Anthropology on a Warming Planet

Climate litigation clearly has implications beyond the courtroom. Having explored these implications in relation to power and politics at a global scale, I now turn to my own academic discipline: What conclusions can be drawn from this study for the practice of anthropology? As Eriksen (2006) explained in his book on public engagement in the discipline, since the Second World War many anthropologists have been reluctant to disseminate their insights beyond academia. To this day, the discipline sees little representation in public discussions around the world. As such, anthropology rarely achieves the relevance it deserves, despite its potential to help people understand the world—and to change it. Anthropologists have made little contribution to political discussions about climate change, and anthropological knowledge tends to be sidelined at international institutions such as the IPCC and UNFCCC. In a review of climate change anthropology, Jessica O'Reilly and colleagues (2020) argue that this is a lost opportunity as anthropological approaches can help rethink political solutions for climate change and reimagine human-atmosphere relations. Eriksen (2020) contends that climate change makes it all the more urgent for anthropologists to raise their voices in public discussions. We may not have all the answers, but we do ask some of the right questions. In addition, suggests Eriksen, anthropologists are able to tell compelling stories that can help people envision different futures and convince them to stand up for change.

While anthropological research cannot provide definitive answers to the normative questions underlying global concerns about climate change, it can uncover the values, standpoints, and relationships at stake. Climate change entangles people, nonhuman persons—from corporations to mountains—and material environments in moral bonds. Anthropological analysis can show how people approach these relations from different perspectives, and we as anthropologists can push for open political discussions that account for multiple standpoints. In dialogue with social activists, academic researchers, policy makers, and other interested parties, I hope that anthropologically informed approaches will allow for more informed exchange and decision making.

Anthropology has the potential to trace how people around the world are engaging with changing environments, to unpack the epistemological and normative stakes of scientific climate change knowledge in its production and dissemination, and to question dominant political and economic frameworks. If we as academics are able to communicate these insights with broader public audiences, we might help people recognize how different solutions are possible. In this book, I show how the legal claim between Luciano Lliuya and RWE configured climate change in terms of neighborly relations involving people like Luciano Lliuya, corporations such as RWE, and nonhuman mountain beings. At the end of this analysis, a final question lingers beneath the surface: What about us anthropologists? What kinds of neighbors should we be? What responsibility do we have, and to whom?

One of our primary responsibilities is to the people we study, our interlocutors. Offering us the relational engagements in which our discipline is rooted, we have a moral duty—formalized, though often inadequately, in institutional ethics guidance—to prevent them from encountering harm as a result of our work. And if we conduct research about climate change, we might even be drawn to interlocutors who face the worst impacts. We are also responsible to our students; we not only impart our knowledge to them but urge them to ask critical questions as they engage with the world. In all its disciplinary breadth, one of anthropology's foundational attributes is its capacity to question what people take for granted. As such, we can lend our students the anthropological sensibility to interrogate the failure of our political and economic systems to forestall a looming climate disaster.

Bearing in mind these underlying responsibilities, another fundamental question arises: Why do we produce anthropological knowledge? Eriksen (2006, 16) draws a useful distinction between analytical work and advocacy. Accordingly, anthropology's core disciplinary commitment is to analyze why and how things are the way they are. It does not answer the normative question of how things should be. As such, analytic work in anthropology is of a different nature than advocacy in which anthropologists take a political stand. At times, the two may be difficult to reconcile: While advocacy usually requires a clear commitment to a particular set of ideas, anthropology does its best to pick apart and defamiliarize everything it encounters. Nevertheless, anthropological analysis can certainly inform normative understandings—both our own and those of the people with whom we share our insights.

In this study, I use an anthropological approach to unpack the underlying assumptions and social dynamics emerging in a transnational climate litigation lawsuit. I examine how the claim functioned and reflect on its significance beyond the courtroom. Throughout my research journey, the anthropological lens impelled me to ask ever more questions. It did not provide normative answers to those questions. Analysis potentially undermines advocacy if it points to the potential contradictions inherent in normative commitments. Or, I might hope, an open engagement with the underlying difficulties in one's normative stance could strengthen social and political arguments. The world, as anthropology teaches us, is full of contradictions.

As such, anthropologists studying climate change—one of the decisive issues facing our planet today—should be aware of the implications that their knowledge might have for the world beyond the academic sphere. This sensibility might even inform our research—between planning, fieldwork, analysis, and dissemination—and our engagement with students. Even if anthropological analysis itself does not answer normative questions about climate change, our personal ethical commitments and professional responsibilities should guide our work as anthropologists. And in the context of climate change, these commitments and responsibilities may leave us with no choice but to take a stand; to engage more proactively, for example, with nonacademic audiences.

My engagement in climate litigation began as a professional activist working for an NGO. This engagement changed shape when I became a professional anthropologist conducting ethnographic research. Throughout my academic work, I remain committed to anthropological analysis and climate justice advocacy. From both a scholarly and activist perspective, climate change is too urgent for me to resort to the illusion of academic neutrality. In academia, particularly in the Global North, we write from a position of privilege. Many of us will be insulated from the worst impacts of climate change that linger in the near future—from glacial retreat and water scarcity, or from rising sea levels that swallow entire countries. I can choose whether or not to think about climate change. Luciano Lliuya, as he has told me in many conversations, does not have a choice. In the long run, climate change threatens his community's way of life. If glaciers disappear and rivers run dry, his fields will be barren. He feels the environment suffering and wonders what will become of its sentient inhabitants.

Luciano Lliuya chose to take a stand, collaborating in a legal complaint that has drawn attention around the world. In studying this claim, I seek to intervene in scholarly discussions about climate change. However, taking inspiration from Luciano Lliuya, I also hope to help address the climate emergency. As anthropologists, we have a responsibility to our interlocutors and our students who will inherit this planet. To this, I add a further dimension: If anthropology entails attending to social life in all its complexity, I contend that we also have a disciplinary responsibility to social life itself. Climate change threatens life as we know it, in the most literal sense. Let us be good neighbors on this warming planet.

* * *

LATE AT NIGHT after the court hearing in 2017, I sat with Luciano Lliuya as he packed his suitcase. After celebrating an unexpected legal milestone and giving interviews to media outlets around the world, he was preparing to go home. I asked what the monumental day in court would mean for him when he returned to Peru.

"When I go home," he replied, "the glaciers will continue to melt. With our claim, we hope to make a small contribution so that sometime in the future the big industries will stop polluting the environment and global warming will stop. Hopefully, the glaciers can then find a new balance." He would explain to his family that he and Germanwatch had set a significant legal precedent that would, they hoped, push politicians to action. But he would also talk to Churup, the mountain above his village. His whole life he had lived below the mountain, looking up at its pristine white glaciers. It possesses a powerful presence in the village community. Since he was young, Luciano Lliuya watched the white mountain as it became infected with dark spots. Snow and ice were quickly disappearing. As long as he could remember, this change made him sad and angry. The mountains gave him the motivation to fight for justice in the courts and on the global stage. Finally, he could return to the mountains with something in hand.

"I'll arrive and look up to Mount Churup." His eyes welled and his voice began to break. A single tear rolled down his cheek. "I'll look up to Mount Churup, and I'll say: 'Now I'm back here—and we made it. We made it.'"

Afterword

On May 28, 2025, the Upper State Court in Hamm announced a historic verdict. In the case of *Luciano Lliuya v. RWE*, it found that in principle, major greenhouse gas emitters can be held liable for their contribution to climate harms. However, the lawsuit was dismissed as the court determined that the risk of flooding to Luciano Lliuya's house was too low. The judgment nevertheless marked a turning point: For the first time ever, anywhere in the world, a court established the principle of corporate climate liability. It set a significant precedent that may be replicated by courts worldwide. It gives vulnerable communities a strong tool to hold major emitters accountable. Countless others face even higher risks and worse climate impacts than Luciano Lliuya, and they may soon have their day in court.

In the 139-page verdict, the judges explained that climate change impacts infringe on people's property rights, and those affected have the right to make claims against major polluters. They found that emitters like RWE have made a direct contribution to climate harms and that the processes are well understood scientifically. Attribution science can be used to establish a causal chain linking individual emitters with individual impacts. According to the court, these processes were foreseeable at least as far back as 1965, meaning RWE could be held liable for its contribution to climate change since then. The judgment clarified that only major emitters could be held responsible, rather than individual consumers who make a negligible contribution to climate change. An emissions share as low as 0.25 percent could be considered relevant, the judges found, as this proportion is high in relation to other emitters. Companies could also be held liable for emissions produced by their subsidiaries. Finally, the judges ruled that climate litigation is a legitimate tool for seeking corporate accountability.

The court dismissed the case on the facts, finding that the risk to Luciano Lliuya's house was too low to merit a legal claim. The judges followed the assessment of the court-appointed scientific expert Rolf Katzenbach, who found that the risk of flooding was less than 1 percent over the next thirty years. The assessment was criticized by the plaintiff's experts as Katzenbach did not consider how further climate change may increase the risk. Local warming causes mountains to become less stable, making them more prone to large rock avalanches. Weeks before the ruling, it became known that Katzenbach had had a business relationship with an RWE subsidiary before and during his involvement in the trial. Luciano Lliuya's lawyers filed a motion to remove the expert, but this was denied by the court, which found that the relationship was not substantial enough and that the motion was filed too late. As the case failed over the first evidentiary question of flood risk, the court did not rule on the second question of whether that risk could be attributed to RWE. If the risk had been higher, and the plaintiff had provided sufficient proof to link the risk to global climate change and to RWE's emissions, the company would have been held fully liable.

The verdict was a significant step forward for climate justice. The outcome was formally a dismissal, but Luciano Lliuya and his supporters celebrated it as a historic win. Legal systems are not built for radical change. Historically, judiciaries were established to maintain power structures and enforce existing rules. Strategic litigation uses the law to shift power relations in the present day. Legal change usually occurs incrementally, and by that measure Luciano Lliuya's court decision is a milestone. It clarifies that existing laws can be used to hold major emitters responsible for their contribution to climate change. This ruling sends a powerful message to courts around the world and sets the stage for future verdicts that make polluters pay. One lawsuit will not save the climate, but a critical mass of cases may push the needle in the right direction. Luciano Lliuya's claim against RWE achieved much more than its initiators dared to imagine. The decade-long odyssey through the courts was a grueling effort toward a dream of justice. The judgment is a shift in the tide that offers hope for a better future.

Acknowledgments

The first person to thank is Saúl Luciano Lliuya. Without him, none of this would have happened. For over ten years he has been a partner in activism, an ethnographic informant, but most of all a friend. When I started this research project, I thought I knew what the lawsuit was all about—a legal and political claim for climate justice. Following Luciano Lliuya to the Peruvian Andes, I learned that it concerned much more: from precarious rural livelihoods to powerful mountain beings.

The idea for this book arose out of my work with Germanwatch on the lawsuit between Luciano Lliuya and RWE. This project would not have been possible without support from Klaus Milke, the former chair of Germanwatch, and Roda Verheyen, Luciano Lliuya's lead lawyer, who gave me endless insight into legal and activist processes. I also thank Roxana Baldrich, Christoph Bals, Marlene Becker, Clara Goldmann, Julia Grimm, Lena Hildebrandt, Lukas Kiefer, Mascha Klein, John Peters, and Caroline Schroeder, who all helped this project come to fruition. Will Frank was particularly helpful in providing feedback from a legal perspective; he sadly passed away in 2023 and will be remembered as the person who came up with the idea of applying German nuisance law to climate change.

I conducted twenty months of fieldwork in Peru for this project. I express enormous thanks to the Rosales family, who took me in and showed me what life is like in the rural Andes. Thanks to my *compadre* Carlos and *comadre* Margarita and to Guillermina, Olga, and Antonio, as well as Jimena, Heinz, Yesica, Emiliano, Sergio, and Lizeth. I am especially grateful to Alejandro, who unfortunately passed away just as this book went into publication. I also thank Julio and Juliana for telling me stories about the past, as well as Brandon and Gleisy for telling me about the future. As Palcacocha became one of my main field sites, I owe endless gratitude to Eduardo and Víctor—not only for their trust in hosting me on countless visits to the lake but also for showing me the power of the landscape and its beings. Thank you to Luis, who taught me so much about the mountains: it was a shame to lose you so young.

In Huaraz, I owe an enormous debt of gratitude to Roca, who always stood by me as a friend and showed endless patience in helping me deal with the inevitable difficulties of ethnographic fieldwork. I thank Inés and Orlando for helping me to learn and understand Quechua, as well as César and Mariluz for helping me understand the past and present of glacial retreat. In Lima, I need to thank Ruth and Enrique for their endless support, as well as Willy, Antonio, and Eyner for their friendship and intellectual input. Thanks also go to Alex Luna, who took most of the photographs in this book; on our journeys through the Andes I came to see both the landscape and its people in a new light through the lens of his camera.

Over the past decade, I have received feedback and helpful comments from numerous academics in countless conferences, seminars, and discussions. I completed this book across my stays at several institutions. At Manchester University, my PhD supervisors Penny Harvey and Chika Watanabe helped my ideas flourish with their support and kindness. I also thank Anna Balázs, Elena Borisova, Tom Boyd, Jeanette Edwards, Guilherme Fians, John F. Foster, Jeremy Gunson, Tree Kelly, Steph Meysner, Joana Nascimento, Akimi Ota, Vlad Schüler, Pedro Silva Rocha Lima, Laura Thurmann, Diego Valdivieso, Marisol Verduga, and Jérémie Voirol. Thanks to Sheila Jasanoff at Harvard University. From my time at University College London, I thank Lisa Vanhala, who helped me gain a foothold in postdoctoral life, as well as Frances Butler, Elisa Calliari, Friederike Harz, Angelica Johansson, and Monserrat Madariaga. I owe endless gratitude to Joana Setzer at London School of Economics (LSE), who believed in me and helped me become the best version of myself. I also thank my colleagues Emily Bradeen, Tiffanie Chan, Ian Higham, Kate Higham, Isabella Keuschnigg, Andrea Pia, Nick Petkov, and Joy Reyes. Thanks to Joeri Rogelj, Friederike Otto, and Emily Theokritoff at Imperial College London.

Numerous others provided input, advice, and support at various stages of this book project. Thomas Hylland Eriksen was a source of immense inspiration and guidance—with his passing, anthropology lost one of the greats. Thanks to Rupert Stuart-Smith for helping me understand attribution science. Christian Huggel showed me why climate science matters. I also thank Jahir Anicama, Michael Burger, Mark Carey, Melina de Bona, Fabian Drenkhan, Adam Emmer, Liz Fisher, Laura Gyte, Matthew Keracher, Sandhya Narayanan, Ben Orlove, César Rodríguez-Garavito, Zoe Savitsky, Joe Snape, Letty Thomas, and April Williamson.

Throughout my career path I have counted on help from Jamie and David, who taught me the joy of reading and writing. I am endlessly grateful for all the support they have given me over the years to make me the person I am today. I also need to thank Konstantin and Alex for their friendship and support. I thank Carolina for always believing in me and for reminding me about the joys of life beyond my work. Finally, I thank Matilda for reminding me not to take everything so seriously.

Parts of the introduction and chapter 2 appeared in an article in *Anthropological Theory* titled "Climate Change in the Courtroom: An Anthropology of Neighborly Relations." An earlier version of chapter 3 appeared in an article in the *Journal of Legal Anthropology* titled "Justice for Glaciers: The Politics of Personhood in Climate Change Litigation." Some portions of part 2 were featured in an article in *Transnational Environmental Law* titled "Save the Climate but Don't Blame Us: Corporate Arguments in Climate Litigation."

Thank you to Gisela Fosado at Duke University Press for believing in this project from the time when it was just an idea, and helping it come to life. Alejandra Mejía provided invaluable support throughout the book's production process. This research was supported by the Economic and Social Research Council of the United Kingdom (Grant references: ES/W00674X/1, ES/Y003314/1).

Notes

INTRODUCTION: CLIMATE JUSTICE IN COURT

1. All quotations from Peruvians are translated from Spanish by the author, unless otherwise noted.

2. “Las lagunas son las lágrimas de las montañas. Hoy, la justicia escuchó las montañas llorando.”

3. Most notably, *Native Village of Kivalina v. ExxonMobil Corporation, et al.*, a case brought by an Alaska community against US energy companies concerning the rising sea level, was dismissed in 2009 (Belleville and Kennedy 2012).

4. Courtroom dialogue is translated from German by the author.

5. At the time of filing, the claim stated that RWE was responsible for 0.47 percent of industrial greenhouse gas emissions. This figure was amended to 0.38 percent in March 2025 based on updated historical emissions data. The revised figure was used in the final stages of the lawsuit.

6. While some distinguish between morality and ethics, I use the terms interchangeably. See Mattingly and Throop 2018.

1. GLACIERS MELT INTO THE COURTROOM

1. The word Palcacocha is Quechua. *Cocha* means lake and *palca* means bifurcation, referring to the shape of the mountain above the lake.

2. The Fifth Assessment Report of the Intergovernmental Panel on Climate Change (IPCC) highlights glacial retreat in the Cordillera Blanca as a significant climate change impact (IPCC 2014, 1519). Environmental research points to the link between climate change, glacial retreat, and glacial lake outburst flood risk in the area (Emmer and Cochachin 2013).

3. On the role of subnational litigation for regulating greenhouse gas emissions, see Osofsky 2007b; on the relation between climate change and tort law, see Kysar 2011; and on the potential for raising public awareness through litigation, see Rogers 2013.

4. In the natural scientific literature on flooding, authors typically differentiate between hazard and risk: while *hazard* denotes the physical threat of an event such as an avalanche or outburst flood, *risk* arises when a hazard threatens particular values such as human life or property. In this sense, risk is a combination of physical hazard, values at stake, and the likelihood for harm to occur (Kron 2005).

5. Rasmussen (2015) has made similar observations in his research among rural communities in the southern Cordillera Blanca.

6. Allen (1988) and de la Cadena (2015) are significant examples.

2. DAVID AND GOLIATH IN THE COURTROOM

1. For the full database, see Climate Change Litigation Databases, accessed June 25, 2025, http://climatecasechart.com/. A majority of documented cases are in the United States.

2. Rooted in the tradition of civil law, court rulings in Germany usually provide normative guidance for future rulings rather than being strictly legally binding as in most common law systems (von Ungern-Sternberg 2013). Nevertheless, the judgment in *Luciano Lliuya v. RWE* constitutes a precedent in a broader sense: Numerous other countries have laws similar to those applied in the German case, meaning that the plaintiff's legal argumentation can be adapted for similar cases in other jurisdictions.

3. The court was unable to identify a scientific consensus for determining a specific emissions reduction pathway for the defendant. Milieudefensie v. Shell Plc, The Hague Court of Appeal (The Netherlands), 200.302.332/01, Judgment of 12 Nov 2024.

4. Saúl Ananías Luciano Lliuya v. RWE AG, Landgericht Essen, Az.: 2 O 285/15, lawsuit of Luciano Lliuya, November 24, 2015. All quotes from legal documents in *Luciano Lliuya v. RWE* are the author's translation from German.

5. For a summary of the flood-modeling study, see Somos-Valenzuela et al. (2016).

6. Authorized translation from the German Ministry of Justice and Consumer Protection (Bundesamt für Justiz 2013b).

7. Reichsgericht in Zivilsachen (Reich Civil Court) (rgz) 167, 14, 24.

8. Saúl Ananías Luciano Lliuya v. RWE AG, legal brief of Luciano Lliuya, July 11, 2016, 15.

9. Saúl Ananías Luciano Lliuya v. RWE AG, legal brief of RWE, April 28, 2016, 42–43.

10. In German law, *immission* refers to the effects on an incoming emission to a property or living organism in terms of air, ground, or water pollution. See Ule et al. 2014.

11. Author's translation from German.

12. In the German civil legal system, appeals courts such as the Upper State Court in Hamm (*Oberlandesgericht*) do not merely review the legal reasoning of lower courts. They are empowered to reopen the factual record, admit new evidence, and oversee expert testimony. In *Luciano Lliuya v. RWE*, the State Court in Essen (*Landg ericht*) dismissed the claim as inadmissible (*unzulässig*), effectively ending the case before it reached the evidentiary stage. This is roughly equivalent to a pretrial dismissal in common law systems such as the United States. On appeal, the Hamm court found the claim admissible and initiated a full evidentiary process (*Beweisaufnahme*), including the commissioning of expert reports and a site visit to Peru. While technically an appeal, this stage functioned much like a trial in common law systems.

13. Formally, the Stiftung Zukunftsfähigkeit (Foundation for Sustainability) made this commitment. The foundation is associated with Germanwatch and provides the NGO with financial backing.

14. In the local (*Callejón de Huaylas*) version of Quechua, neighbors are usually referred to as *markamayi* (person from the same village) or *wayii waknichaw taraq* (person who lives next to my house).

15. Wilhelm Frank was part of Luciano Lliuya's legal team until he passed away in 2023.

INTERLUDE 1: ANDEAN LIFE IN AN UNCERTAIN CLIMATE

1. Name changed.

2. Name changed.

3. THE POLITICS OF PERSONHOOD

Note: An earlier version of chapter 3 appeared as the article "Justice for Glaciers: The Politics of Personhood in Climate Change Litigation," in *Journal of Legal Anthropology* 8, no. 1 (2024).

1. Anthropologists have documented the construction of personhood in numerous contexts since Marcel Mauss argued in a 1938 lecture that personhood is a social phenomenon (Mauss 1985). For an overview, see Carsten 2004.

2. Citizens United v. Federal Election Commission, 558 U.S. 310 (2010).

PART II: CAUSALITY IN THE COURTROOM

Epigraphs: City and County of Honolulu v. Sunoco LP, Circuit Court of the First Circuit of the State of Hawaii, 1CCV-20-0000380, Brief of Chevron Corporation and Chevron U.S.A. Inc., September 12, 2022, 3; Milieudefensie et al. v. Royal Dutch Shell PLC, The Hague District Court C/09/571932 2019/379, Brief of Royal Dutch Shell, November 13, 2019, 10. I reference an unofficial English translation of the Dutch original from the Global Climate Change Litigation Database of the Sabin Center for Climate Change Law at Columbia Law School, https://climatecasechart.com/non-us-case/milieudefensie-et-al-v-royal-dutch-shell-plc/.

4. TRUTH AND RESPONSIBILITY IN THE COURTROOM

1. This is my summary; for a full account of the evidentiary questions, see the court's ruling on the matter. Saúl Ananías Luciano Lliuya v. RWE AG, Oberlandesgericht Hamm, Az.: I-5 U 15/17, ruling from September 27, 2021.

2. Saúl Ananías Luciano Lliuya v. RWE AG, Oberlandesgericht Hamm, Az.: I-5 U 15/17, ruling from July 1, 2021, 6. In the court's final judgement, it settled on 1965 for the time of reasonable foreseeability (Saúl Ananías Luciano Lliuya v. RWE AG, Oberlandesgericht Hamm, Az.: I-5 U 15/17, Judgment from May 28, 2025, 47).

3. Saúl Ananías Luciano Lliuya v. RWE AG, Landgericht Essen, Az.: 2 O 285/15, lawsuit of Luciano Lliuya, November 24, 2015, 2.

4. *Conditio sine qua non* is Latin and refers to "a condition without which the harm would not have occurred." Verheyen 2015, 163.

5. Legally established facts may be subsequently disputed or revised during appeals or other related proceedings.

6. Bundesgerichtshof (Federal Court of Justice) ruling from 17.02.1970 (III ZR 139/67, Anastasia Decision), translated by author. This decision emerged from the case of a woman who claimed to be Anastasia, daughter of the last Russian Czar, who was reportedly murdered with her family in the aftermath of the Russian Revolution. The woman in question claimed that she had secretly escaped Russia and settled in Germany, where she sought to acquire a portion of the Czar family's inheritance. Judges dismissed the case on the grounds that she had not provided sufficient evidence to prove her alleged identity.

5. TRACING EMISSIONS

1. While the judges would decide whom to appoint as expert witnesses, they sought advice from the parties to the suit. After the two sides in this case failed to agree which scientists were best qualified, the judges independently sought out and appointed a set of scientific experts.

2. The court chose Rolf Katzenbach (Technical University Darmstadt) and Johannes Hübl (Universität für Bodenkultur Wien [BOKU], University of Natural Resources and Life Sciences, Vienna).

3. As I explained in chapter 4, cumulative causality exists when multiple parties have contributed to cause the same process or event.

4. These cases were summarily addressed in a 1987 ruling of the Federal Court of Justice. Waldschadensurteil, BGH (Bundesgerichtshof), 10.12.1987—III ZR 220/86.

5. Saúl Ananías Luciano Lliuya v. RWE AG, Landgericht Essen, Az.: 2 O 285/15, legal brief of RWE, April 28, 2016, 38–40.

6. Saúl Ananías Luciano Lliuya v. RWE AG, legal brief of Luciano Lliuya, July 11, 2016, 13–14.

7. Roda Verheyen and Peter Roderick at the Climate Justice Programme first commissioned the research with Richard Heede in 2003. Before this, only country-specific emissions data were available. The study's results have been published in peer-reviewed journals. See Heede 2014b; Frumhoff et al. 2015; Ekwurzel et al. 2017.

8. The 0.47 percent figure comes from the original Carbon Majors dataset (1854–2010). An update to the Carbon Majors database released in 2025 recalculated RWE's legacy at approximately 0.38 percent through 2023 (https://carbonmajors.org/Entity/RWE-136).

9. Author's translation from German.

10. This figure was updated to 0.38 percent in 2025 shortly before the case's conclusion.

11. Saúl Ananías Luciano Lliuya v. RWE AG, legal brief of RWE, April 28, 2016, 25–26.

12. Saúl Ananías Luciano Lliuya v. RWE AG, legal brief of RWE, April 28, 2016, 24.

13. Saúl Ananías Luciano Lliuya v. RWE AG, lawsuit of Luciano Lliuya, November 24, 2015, 11.

14. Greenhouse Gas Emissions Trading Act (Treibhausgas-Emissionshandelsgesetz).

15. Saúl Ananías Luciano Lliuya v. RWE AG, lawsuit of Luciano Lliuya, November 24, 2015, 12.

16. Saúl Ananías Luciano Lliuya v. RWE AG, legal brief of RWE, April 28, 2016.

17. Saúl Ananías Luciano Lliuya v. RWE AG, legal brief of RWE, April 28, 2016, 37.

6. MODELING THE GLOBAL CLIMATE

1. Saúl Ananías Luciano Lliuya v. RWE AG, Landgericht Essen, Az.: 2 O 285/15, lawsuit of Luciano Lliuya, November 24, 2015, 13, 35–36.

2. Saúl Ananías Luciano Lliuya v. RWE AG, lawsuit of Luciano Lliuya, 16.

3. Saúl Ananías Luciano Lliuya v. RWE AG, Landgericht Essen, Az.: 2 O 285/15, legal brief of RWE, April 28, 2016, 13, 15–21.

4. Saúl Ananías Luciano Lliuya v. RWE AG, legal brief of RWE, 13.

5. Saúl Ananías Luciano Lliuya v. RWE AG, Oberlandesgericht Hamm, Az.: I-5 U 15/17, legal brief of Luciano Lliuya, April 15, 2021.

6. Saúl Ananías Luciano Lliuya v. RWE AG, Oberlandesgericht Hamm, Az.: I-5 U 15/17, legal brief of RWE, December 15, 2021, 5.

7. These were Somos-Valenzuela et al. (2016) and Frey et al. (2018).

8. Saúl Ananías Luciano Lliuya v. RWE AG, legal brief of RWE, March 22, 2019.

9. The report addressed only the conclusions by Stuart-Smith et al. (2021b) regarding flood risk, not the study's primary focus, which was the attribution of local temperature increase and glacial retreat to global warming. RWE's lawyers did not reveal who advised them on climate and attribution science.

10. Saúl Ananías Luciano Lliuya v. RWE AG, legal brief of RWE, December 15, 2021, 19, 23.

11. Saúl Ananías Luciano Lliuya v. RWE AG, Oberlandesgericht Hamm, Az.: I-5 U 15/17, ruling from July 1, 2021.

12. Saúl Ananías Luciano Lliuya v. RWE AG, Oberlandesgericht Hamm, Az.: I-5 U 15/17, legal brief of RWE, December 15, 2021, 23–25.

13. Saúl Ananías Luciano Lliuya v. RWE AG, legal brief of RWE, December 15, 2021, 23.

14. Saúl Ananías Luciano Lliuya v. RWE AG, Oberlandesgericht Hamm, Az.: I-5 U 15/17, legal brief of RWE, April 16, 2018, 10.

15. 947 F.3d 1159 (9th Cir. 2020).

7. MEASURING PALCACOCHA

1. Name changed.

2. Name changed.

3. Known as the Center for Water and the Environment as of 2025.

4. Decreto Supremo 002-2011-PCM.

5. Saúl Ananías Luciano Lliuya v. RWE AG, Landgericht Essen, Az.: 2 O 285/15, legal brief of RWE, April 28, 2016, 33–34.

6. Saúl Ananías Luciano Lliuya v. RWE AG, Oberlandesgericht Hamm, Az.: I-5 U 15/17, legal brief of RWE, October 20, 2017, 18–19.

7. Saúl Ananías Luciano Lliuya v. RWE AG, Oberlandesgericht Hamm, Az.: I-5 U 15/17, legal brief of RWE, December 15, 2021, 12.

8. "Article Processing Charges (APC) Information," MDPI, accessed June 29, 2025, https://www.mdpi.com/apc.

9. The article, entitled "On the Misdiagnosis of Surface Temperature Feedbacks from Variations in Earth's Radiant Energy Balance" (Spencer and Braswell 2011),

remains on the journal's website as of 2025 with the note that "this paper attracts great attention."

10. "Instructions for Authors," *Remote Sensing*, MDPI, accessed June 29, 2025, https://www.mdpi.com/journal/remotesensing/instructions.

11. Saúl Ananías Luciano Lliuya v. RWE AG, Oberlandesgericht Hamm, Az.: I-5 U 15/17, Judgment, May 28, 2025, 54.

8. GLACIAL POLITICS

1. Name changed.

2. For example, studies conducted in the Cordillera Blanca have examined GLOF hazard assessment (Schneider et al. 2014; Emmer et al. 2016; Frey et al. 2018; Mergili et al. 2020), GLOF hazard mitigation strategies (Frey et al. 2014; Emmer et al. 2018), glacial retreat (Schauwecker et al. 2014; Emmer et al. 2015), climate change impact attribution (Huggel et al. 2020; Stuart-Smith et al. 2021b), socio-hydrological change (Bury et al. 2013; Carey et al. 2014; Drenkhan et al. 2015; Mark et al. 2017), and engagements with local populations about climate change (Huggel et al. 2015; Jurt et al. 2015).

3. To facilitate greater institutional coordination on climate-related hazards, the Province of Huaraz and District of Independencia formed the Mancomunidad Municipal de Waraq (Waraq Municipal Association) in 2013, which played a significant role in pushing for an early-warning system at Palcacocha.

9. ENGINEERING IN A SENTIENT ENVIRONMENT

1. Words drawn from a subsequent interview with Díaz, when I asked him to recite what he had said during the pago. Translation from Quechua by Inés Yanac León.

2. This argument builds on Isabel Stengers's "cosmopolitical proposal," which calls for us to account for forms of existence in political processes beyond the human-centered understandings common in political theory (Stengers 2005).

CONCLUSION: CHANGING THE LEGAL CLIMATE

1. For a case study from Paraguay, see Hetherington 2013.

References

Allen, Catherine J. 1988. *The Hold Life Has: Coca and Cultural Identity in an Andean Community*. Smithsonian Series in Ethnographic Inquiry. Washington, DC: Smithsonian Institution Press.

Allen, Catherine J. 1997. “When Pebbles Move Mountains: Iconicity and Symbolism in Quechua Ritual.” In *Creating Context in Andean Cultures*, edited by Rosaleen Howard-Malverde, 73–84. New York: Oxford University Press.

Alogna, Ivano, Christine Bakker, and Jean-Pierre Gauci. 2021. *Climate Change Litigation: Global Perspectives*. Leiden, Netherlands: Brill, Nijhoff.

Appel, Hannah, Nikhil Anand, and Akhil Gupta. 2018. “Introduction: Temporality, Politics, and the Promise of Infrastructure.” In *The Promise of Infrastructure*, edited by Nikhil Anand, Akhil Gupta, and Hannah Appel, 1–38. Durham, NC: Duke University Press.

Appiah, Kwame Anthony. 2005. *The Ethics of Identity*. Princeton, NJ: Princeton University Press.

Åtland, Kristian. 2010. “Russia and Its Neighbors: Military Power, Security Politics, and Interstate Relations in the Post–Cold War Arctic.” *Arctic Review* 1 (2): 279–98.

Aust, Helmut Philipp. 2022. “Climate Protection Act Case, Order of the First Senate.” *American Journal of International Law* 116 (1): 150–57.

Auz, Juan. 2022a. “Human Rights–Based Climate Litigation: A Latin American Cartography.” *Journal of Human Rights and the Environment* 16 (1): 114–36.

Auz, Juan. 2022b. “Two Reputed Allies.” In Rodríguez-Garavito, *Litigating the Climate Emergency*, 145–56.

Barnes, Jessica, Michael Dove, Myanna Lahsen, Andrew Mathews, Pamela McElwee, Roderick McIntosh, Frances Moore, Jessica O’Reilly, Ben Orlove, and Rajindra Puri. 2013. “Contribution of Anthropology to the Study of Climate Change.” *Nature Climate Change* 3 (6): 541–44.

Bashkow, Ira. 2014. “Afterword: What Kind of a Person Is the Corporation?” *PoLAR: Political and Legal Anthropology Review* 37 (2): 296–307. https://doi.org/10.1111/Plar.12076.

Beauregard, Charles, D’Arcy Carlson, Stacy-ann Robinson, Charles Cobb, and Mykela Patton. 2021. “Climate Justice and Rights-Based Litigation in a Post-Paris World.” *Climate Policy* 21 (5): 652–65.

Beck, Ulrich. 2006. *Cosmopolitan Vision*. Cambridge: Polity Press.

Belleville, Mark, and Katherine Kennedy. 2012. "Cool Lawsuits—Is Climate Change Litigation Dead After *Kivalina v. Exxonmobil*?" *Appalachian Natural Resources Law Journal* 7: 51–86. https://heinonline.org/HOL/P?h=hein.journals/anrlj7&i=57.

Bens, Jonas, and Larissa Vetters. 2018. "Ethnographic Legal Studies: Reconnecting Anthropological and Sociological Traditions." *Journal of Legal Pluralism and Unofficial Law* 50 (3): 239–54. https://doi.org/10.1080/07329113.2018.1559487.

Berwyn, Bob. 2021. "For a City Staring down the Barrel of a Climate-Driven Flood, a New Study Could Be the Smoking Gun." *Inside Climate News*, February 4. https://insideclimatenews.org/news/04022021/for-a-city-staring-down-the-barrel-of-a-climate-driven-flood-a-new-study-could-be-the-smoking-gun/.

Blair, Margaret M. 2013. "Corporate Personhood and the Corporate Persona Symposium: In the Boardroom." *University of Illinois Law Review* 2013 (3): 785–820. https://heinonline.org/HOL/P?h=hein.journals/unilllr2013&i=810.

Blumberg, Phillip I. 1990. "The Corporate Personality in American Law: A Summary Review." *American Journal of Comparative Law* 38: 49–69.

Bode, Barbara. 1989. *No Bells to Toll*. New York: Charles Scribner's Sons.

Boehm, Sophie, Louise Jeffery, Kelly Levin, Judit Hecke, Clea Schumer, Claire Fyson, Aman Majid, Joel Jaeger, Anna Nilsson, and Stephen Naimoli. 2022. *State of Climate Action 2022*. Berlin and Cologne, Germany; San Francisco; and Washington, DC: Bezos Earth Fund, Climate Action Tracker, Climate Analytics, ClimateWorks Foundation, NewClimate Institute, United Nations Climate Change High-Level Champions, and World Resources Institute.

Boelens, Rutgerd. 2014. "Cultural Politics and the Hydrosocial Cycle: Water, Power, and Identity in the Andean Highlands." *Geoforum* 57: 234–47. http://www.sciencedirect.com/science/article/pii/S0016718513000432.

Bold, Rosalyn. 2019. "Introduction: Creating a Cosmopolitics of Climate Change." In *Indigenous Perceptions of the End of the World: Creating a Cosmopolitics of Change*, edited by Rosalyn Bold, 1–28. Cham, Switzerland: Palgrave Macmillan.

Bolin, Inge. 2009. "The Glaciers of the Andes Are Melting: Indigenous and Anthropological Knowledge Merge in Restoring Water Resources." In *Anthropology and Climate Change: From Encounters to Actions*, edited by Susan Alexandra Crate and Mark Nuttall, 228–39. Walnut Creek, CA: Left Coast Press.

Boom, Keely. 2016. *Climate Justice: The International Momentum Towards Climate Litigation*. Sydney: Climate Justice Programme.

Boutros, Magda. 2022. "Legal Mobilization and Branches of Law: Contesting Racialized Policing in French Courts." *Law and Society Review* 56 (4): 623–45.

Bouwer, Kim. 2020. "Lessons from a Distorted Metaphor: The Holy Grail of Climate Litigation." *Transnational Environmental Law* 9 (2): 347–78. https://www.cambridge.org/core/article/lessons-from-a-distorted-metaphor-the-holy-grail-of-climate-litigation/40B0DC6E8F3A54AA2A9B4908DFA7E46F.

Bouwer, Kim, Uzuazo Etemire, Tracy-Lynn Field, and Ademola Oluborode Jegede. 2024. *Climate Litigation and Justice in Africa*. Bristol, UK: Bristol University Press.

Brulle, Robert J. 2014. "Institutionalizing Delay: Foundation Funding and the Creation of U.S. Climate Change Counter-Movement Organizations." *Climatic Change* 122 (4): 681–94. https://doi.org/10.1007/s10584-013-1018-7.

Bundesamt für Justiz. 2013a. "Code of Civil Procedure." Bundesministerium der Justiz und für Verbraucherschutz. Accessed November 16, 2020. https://www.gesetze-im-internet.de/englisch_bgb/englisch_bgb.html.

Bundesamt für Justiz. 2013b. "German Civil Code." Bundesministerium der Justiz und für Verbraucherschutz. Accessed November 16, 2020. https://www.gesetze-im-internet.de/englisch_bgb/englisch_bgb.html.

Burger, Michael, Jessica Wentz, and Radley Horton. 2020. "The Law and Science of Climate Change Attribution." *Columbia Journal of Environmental Law* 45 (1): 57–240.

Burgers, Laura. 2020. "Should Judges Make Climate Change Law?" *Transnational Environmental Law* 9 (1): 55–75.

Bury, Jeffrey, Bryan G. Mark, Mark Carey, Kenneth R. Young, Jeffrey M. McKenzie, Michel Baraer, Adam French, and Molly H. Polk. 2013. "New Geographies of Water and Climate Change in Peru: Coupled Natural and Social Transformations in the Santa River Watershed." *Annals of the Association of American Geographers* 103 (2): 363–74. http://dx.doi.org/10.1080/00045608.2013.754665.

Callison, Candis. 2014. *How Climate Change Comes to Matter: The Communal Life of Facts*. Durham, NC: Duke University Press.

Cann, Heather W., and Leigh Raymond. 2018. "Does Climate Denialism Still Matter? The Prevalence of Alternative Frames in Opposition to Climate Policy." *Environmental Politics* 27 (3): 433–54. https://doi.org/10.1080/09644016.2018.1439353.

Carey, Mark. 2010. *In the Shadow of Melting Glaciers: Climate Change and Andean Society*. New York: Oxford University Press.

Carey, Mark, Michel Baraer, Bryan G. Mark, Adam French, Jeffrey Bury, Kenneth R. Young, and Jeffrey M. McKenzie. 2014. "Toward Hydro-Social Modeling: Merging Human Variables and the Social Sciences with Climate-Glacier Runoff Models (Santa River, Peru)." *Journal of Hydrology* 518: 60–70. http://www.sciencedirect.com/science/article/pii/S0022169413008159.

Carsten, Janet. 2004. *After Kinship*. Cambridge: Cambridge University Press.

Cavedon-Capdeville, Fernanda de Salles, María Valeria Berros, Humberto Filpi, and Paola Villavicencio-Calzadilla. 2024. "An Ecocentric Perspective on Climate Litigation: Lessons from Latin America." *Journal of Human Rights Practice* 16 (1): 89–106.

Chao, Sophie. 2018. "In the Shadow of the Palm: Dispersed Ontologies Among Marind, West Papua." *Cultural Anthropology* 33 (4): 621–49. https://doi.org/10.14506/ca33.4.08.

Choy, Timothy K. 2011. *Ecologies of Comparison: An Ethnography of Endangerment in Hong Kong*. Experimental Futures: Technological Lives, Scientific Arts, Anthropological Voices. Durham, NC: Duke University Press.

ClientEarth. n.d. "The Greenwashing Files." Accessed August 4, 2022. https://www.clientearth.org/projects/the-greenwashing-files/.

Cochachin Rapre, Alejo, and César Salazar Checa. 2016. Plan Batimétrico de la Laguna Palcacocha. Huaraz: Autoridad Nacional del Agua.

Cole, Simon A., and Alyse Bertenthal. 2017. "Science, Technology, Society, and Law." *Annual Review of Law and Social Science* 13 (1): 351–71. https://doi.org/10.1146/annurev-lawsocsci-110316-113550.

Contraloría General de la República, La. 2017. Nota de Prensa 52-2017-CG/COM. Lima.

Coombes, Brad. 2020. "Nature's Rights as Indigenous Rights? Mis/recognition Through Personhood for Te Urewera." *Espace populations sociétés / Space populations societies*, nos. 1–2.

Crate, Susan A. 2011. "Climate and Culture: Anthropology in the Era of Contemporary Climate Change." *Annual Review of Anthropology* 40 (1): 175–94. http://www.annualreviews.org/doi/pdf/10.1146/annurev.anthro.012809.104925.

Cruikshank, Julie. 2005. *Do Glaciers Listen? Local Knowledge, Colonial Encounters, and Social Imagination*. Brenda and David McLean Canadian Studies series. Vancouver: University of British Columbia Press; Seattle: University of Washington Press.

De la Cadena, Marisol. 2010. "Indigenous Cosmopolitics in the Andes: Conceptual Reflections Beyond 'Politics.'" *Cultural Anthropology* 25 (2): 334–70. https://doi.org/10.1111/j.1548-1360.2010.01061.x.

De la Cadena, Marisol. 2015. *Earth Beings: Ecologies of Practice Across Andean Worlds*. Durham, NC: Duke University Press.

Demeritt, David. 2001. "The Construction of Global Warming and the Politics of Science." *Annals of the Association of American Geographers* 91 (2): 307–37. https://doi.org/10.1111/0004-5608.00245.

Donger, Elizabeth. 2022. "Children and Youth in Strategic Climate Litigation: Advancing Rights Through Legal Argument and Legal Mobilization." *Transnational Environmental Law* 11 (2): 263–89.

Drenkhan, Fabian, Mark Carey, Christian Huggel, Jochen Seidel, and María Teresa Oré. 2015. "The Changing Water Cycle: Climatic and Socioeconomic Drivers of Water-Related Changes in the Andes of Peru." *Wiley Interdisciplinary Reviews: Water* 2 (6): 715–33.

Drenkhan, Fabian, Christian Huggel, Lucía Guardamino, and Wilfried Haeberli. 2019. "Managing Risks and Future Options from New Lakes in the Deglaciating Andes of Peru: The Example of the Vilcanota-Urubamba Basin." *Science of the Total Environment* 665: 465–83. http://www.sciencedirect.com/science/article/pii/S0048969719305522.

Eckert, Julia, and Laura Knöpfel. 2020. "Legal Responsibility in an Entangled World." *Journal of Legal Anthropology* 4 (2): 1–16. https://www.berghahnjournals.com/view/journals/jla/4/2/jla040201.xml.

Eckert, Julia, Zerrin Özlem Biner, Brian Donahoe, and Christian Strümpell. 2012. "Introduction: Law's Travels and Transformations." In *Law Against the State: Ethnographic Forays into Law's Transformations*, edited by Julia Eckert, Brian Donahoe, Christian Strümpell, and Zerrin Özlem Biner, 1–22. Cambridge: Cambridge University Press.

Edwards, Paul N. 2003. "Infrastructure and Modernity: Force, Time, and Social Organization in the History of Sociotechnical Systems." In *Modernity and Technology*, edited by Thomas J. Misa, Philip Brey, and Andrew Feenberg, 185–225. Cambridge, MA: MIT Press.

Ekardt, Felix, and Katharine Heyl. 2022. "The German Constitutional Verdict Is a Landmark in Climate Litigation." *Nature Climate Change* 12 (8): 697–99.

Ekwurzel, B., J. Boneham, M. W. Dalton, R. Heede, R. J. Mera, M. R. Allen, and P. C. Frumhoff. 2017. "The Rise in Global Atmospheric Co2, Surface Temperature, and Sea Level from Emissions Traced to Major Carbon Producers." *Climatic Change* 144 (4): 579–90. https://doi.org/10.1007/s10584-017-1978-0.

Emmer, Adam, and Alejo Cochachin. 2013. "The Causes and Mechanisms of Moraine-Dammed Lake Failures in the Cordillera Blanca, North American Cordillera, and Himalayas." *AUC Geographica* 48 (2): 5–15.

Emmer, Adam, Edwin C. Loarte, Jan Klimeš, and Vít Vilímek. 2015. "Recent Evolution and Degradation of the Bent Jatunraju Glacier (Cordillera Blanca, Peru)." *Geomorphology* 228: 345–55.

Emmer, Adam, Vít Vilímek, Christian Huggel, Jan Klimeš, and Yvonne Schaub. 2016. "Limits and Challenges to Compiling and Developing a Database of Glacial Lake Outburst Floods." *Landslides* 13: 1579–84. http://dx.doi.org/10.1007/s10346-016-0686-6.

Emmer, Adam, Vít Vilímek, and Marco Zapata Luyo. 2018. "Hazard Mitigation of Glacial Lake Outburst Floods in the Cordillera Blanca (Peru): The Effectiveness of Remedial Works." *Journal of Flood Risk Management* 11: S489–S501. http://dx.doi.org/10.1111/jfr3.12241.

Eriksen, Thomas Hylland. 2006. *Engaging Anthropology: The Case for a Public Presence*. Oxford: Berg.

Eriksen, Thomas Hylland. 2016. *Overheating: An Anthropology of Accelerated Change*. London: Pluto Press.

Eriksen, Thomas Hylland. 2020. "A Better Impact Factor: Anthropology and Climate Change." *Anthropology Today* 36 (1): 1–2. https://doi.org/10.1111/1467-8322.12548.

Fassin, Didier. 2011. "A Contribution to the Critique of Moral Reason." *Anthropological Theory* 11 (4): 481–91. https://doi.org/10.1177/1463499611429901.

Fassin, Didier. 2012. "Introduction: Toward a Critical Moral Anthropology." In *A Companion to Moral Anthropology*, edited by Didier Fassin, 1–17. Oxford: Wiley-Blackwell.

Fischer-Lescano, Andreas. 2020. "Nature as a Legal Person: Proxy Constellations in Law." *Law and Literature* 32 (2): 237–62. https://doi.org/10.1080/1535685X.2020.1763596.

Fisher, Elizabeth. 2013. "Climate Change Litigation, Obsession, and Expertise: Reflecting on the Scholarly Response to *Massachusetts v. EPA*." *Law and Policy* 35 (3): 236–60.

Fisher, Elizabeth, Eloise Scotford, and Emily Barritt. 2017. "The Legally Disruptive Nature of Climate Change." *Modern Law Review* 80 (2): 173–201. https://onlinelibrary.wiley.com/doi/abs/10.1111/1468-2230.12251.

Fitz-Henry, Erin. 2018. "Challenging Corporate 'Personhood': Energy Companies and the 'Rights' of Non-Humans." *PoLAR: Political and Legal Anthropology Review* 41 (S1): 85–102. https://doi.org/10.1111/plar.12255.

Frank, Will. 2010. "Climate Change Litigation—Klimawandel und haftungsrechtliche Risiken" (Climate Change Litigation—Climate Change and Legal Liability Risks). *Neue Juristische Wochenschrift: NJW* 63 (51): 3691–92.

Frank, Will, Christoph Bals, and Julia Grimm. 2019. "The Case of Huaraz: First Climate Lawsuit on Loss and Damage Against an Energy Company Before German

Courts." In *Loss and Damage from Climate Change: Concepts, Methods and Policy Options*, edited by Reinhard Mechler, Laurens M. Bouwer, Thomas Schinko, Swenja Surminski, and JoAnne Linnerooth-Bayer, 475–482. Cham, Switzerland: Springer International.

Franta, Benjamin. 2021. "Early Oil Industry Disinformation on Global Warming." *Environmental Politics* 30 (4): 663–68. https://doi.org/10.1080/09644016.2020.1863703.

Fraser, Nancy. 2009. *Scales of Justice: Reimagining Political Space in a Globalizing World*. New York: Columbia University Press.

Frey, H., J. García-Hernández, C. Huggel, D. Schneider, M. Rohrer, C. Gonzales Alfaro, R. Muñoz Asmat, K. Price Rios, L. Meza Román, A. Cochachin Rapre, and P. Masias Chacon. 2014. "An Early Warning System for Lake Outburst Floods of the Laguna 513, Cordillera Blanca, Peru." Analysis and Management of Changing Risks for Natural Hazards, Padua, Italy.

Frey, Holger, Christian Huggel, Patrick Baer, Rachel E. Chisolm, Brian McArdell, Alejo Cochachin, and César Portocarrero. 2018. "Multi-Source Glacial Lake Outburst Flood Hazard Assessment and Mapping for Huaraz, Cordillera Blanca, Peru." *Frontiers in Earth Science* 6: 210.

Frumhoff, Peter C., Richard Heede, and Naomi Oreskes. 2015. "The Climate Responsibilities of Industrial Carbon Producers." *Climatic Change* 132 (2): 157–71. https://doi.org/10.1007/s10584-015-1472-5.

Ganguly, Geetanjali, Joana Setzer, and Veerle Heyvaert. 2018. "If at First You Don't Succeed: Suing Corporations for Climate Change." *Oxford Journal of Legal Studies* 38 (4): 841–68.

Geiling, Natahsa. 2019. "City of Oakland v. BP: Testing the Limits of Climate Science in Climate Litigation." *Ecology Law Quarterly* 46: 683–94.

Goodale, Mark. 2017. *Anthropology and Law: A Critical Introduction*. New York: New York University Press.

Gribetz, Jonathan Marc. 2014. *Defining Neighbors: Religion, Race, and the Early Zionist-Arab Encounter*. Princeton, NJ: Princeton University Press.

Gutmann, Andreas. 2021. *Hybride Rechtssubjektivität: Die Rechte der "Natur oder Pacha Mama" in der ecuadorianischen Verfassung von 2008*. Beiträge zum ausländischen öffentlichen Recht und Völkerrecht 307. Baden-Baden: Nomos Verlagsgesellschaft.

Haack, Susan. 2014. *Evidence Matters: Science, Proof, and Truth in the Law*. New York: Cambridge University Press.

Harvey, David. 2005. *A Brief History of Neoliberalism*. New York: Oxford University Press.

Harvey, David. 2009. *Cosmopolitanism and the Geographies of Freedom*. New York: Columbia University Press.

Harvey, Penelope. 2001. "Landscape and Commerce: Creating Contexts for the Exercise of Power." In *Contested Landscapes: Movement, Exile, and Place*, edited by Barbara Bender and Margot Winer, 197–210. Oxford: Berg.

Harvey, Penelope, and Hannah Knox. 2015. *Roads: An Anthropology of Infrastructure and Expertise*. Ithaca, NY: Cornell University Press.

Hastrup, Kirsten. 2013. "Anthropological Contributions to the Study of Climate: Past, Present, Future." *Wiley Interdisciplinary Reviews: Climate Change* 4 (4): 269–81.

Hastrup, Kirsten, Peter Elsass, Ralph Grillo, Per Mathiesen, and Robert Paine. 1990. "Anthropological Advocacy: A Contradiction in Terms? [and Comments]." *Current Anthropology* 31 (3): 301–11. http://www.jstor.org/stable/2743631.

Haugestad, Anne K. 2004. "Norwegians as Global Neighbors and Global Citizens." In *Future as Fairness: Ecological Justice and Global Citizenship*, edited by Anne K. Haugestad and J. D. Wulfhorst, 217–40. Amsterdam: Rodopi.

Hayes, Graeme. 2013. "Negotiating Proximity: Expert Testimony and Collective Memory in the Trials of Environmental Activists in France and the United Kingdom." *Law and Policy* 35 (3): 208–35. https://doi.org/10.1111/lapo.12004.

Heede, Richard. 2014a. "Carbon Majors: Accounting for Carbon and Methane Emissions, 1854–2010; Methods and Results Report." Climate Mitigation Services. April 7. http://www.climateaccountability.org/pdf/MRR%209.1%20Apr14R.pdf.

Heede, Richard. 2014b. "Tracing Anthropogenic Carbon Dioxide and Methane Emissions to Fossil Fuel and Cement Producers, 1854–2010." *Climatic Change* 122 (1–2): 229–41. http://dx.doi.org/10.1007/s10584-013-0986-y.

Hegglin, E., and C. Huggel. 2008. "An Integrated Assessment of Vulnerability to Glacial Hazards: A Case Study in the Cordillera Blanca, Peru." *Mountain Research and Development* 28 (3–4): 299–309. https://doi.org/10.1177/14634996221138338.

Hellner, Agnes, and Yaffa Epstein. 2023. "Allocation of Institutional Responsibility for Climate Change Mitigation: Judicial Application of Constitutional Environmental Provisions in the European Climate Cases *Arctic Oil, Neubauer*, and *l'Affaire du siècle*." *Journal of Environmental Law* 35 (2): 207–27.

Henig, David. 2012. "'Knocking on My Neighbour's Door': On Metamorphoses of Sociality in Rural Bosnia." *Critique of Anthropology* 32 (1): 3–19. https://doi.org/10.1177/0308275X11430871.

Hetherington, Kregg. 2013. "Beans Before the Law: Knowledge Practices, Responsibility, and the Paraguayan Soy Boom." *Cultural Anthropology* 28 (1): 65–85. http://dx.doi.org/10.1111/j.1548–1360.2012.01173.x.

High, Mette M., and Jessica M. Smith. 2019. "Introduction: The Ethical Constitution of Energy Dilemmas." *Journal of the Royal Anthropological Institute* 25 (S1): 9–28. https://doi.org/10.1111/1467-9655.13012.

Huggel, Christian, Mark Carey, Adam Emmer, Holger Frey, Noah Walker-Crawford, and Ivo Wallimann-Helmer. 2020. "Anthropogenic Climate Change and Glacier Lake Outburst Flood Risk: Local and Global Drivers and Responsibilities for the Case of Lake Palcacocha, Peru." *Natural Hazards and Earth System Sciences* 20 (8): 2175–93. https://nhess.copernicus.org/articles/20/2175/2020/.

Huggel, Christian, Marlene Scheel, Franziska Albrecht, Norina Andres, Pierluigi Calanca, Christine Jurt, Nikolay Khabarov, Daniel Mira-Salama, Mario Rohrer, Nadine Salzmann, Yamina Silva, Elizabeth Silvestre, Luis Vicuña, and Massimiliano Zappa. 2015. "A Framework for the Science Contribution in Climate Adaptation: Experiences from Science-Policy Processes in the Andes." *Environmental*

Science and Policy 47: 80–94. http://www.sciencedirect.com/science/article/pii/S1462901114002202.

Huggel, Christian, Ivo Wallimann-Helmer, Daithi Stone, and Wolfgang Cramer. 2016. "Reconciling Justice and Attribution Research to Advance Climate Policy." *Nature Climate Change* 6 (10): 901–8. http://dx.doi.org/10.1038/nclimate3104.

Hulme, Mike. 2010a. "Cosmopolitan Climates." *Theory, Culture, and Society* 27 (2–3): 267–76. https://doi.org/10.1177/0263276409358730.

Hulme, Mike. 2010b. "Problems with Making and Governing Global Kinds of Knowledge." *Global Environmental Change* 20 (4): 558–64. http://www.sciencedirect.com/science/article/pii/S0959378010000646.

Hulme, Mike. 2014. "Attributing Weather Extremes to 'Climate Change': A Review." *Progress in Physical Geography: Earth and Environment* 38 (4): 499–511. https://doi.org/10.1177/0309133314538644.

INAIGEM (Instituto Nacional de Investigación en Glaciares y Ecosistemas de Montaña). 2017. "El nivel de agua de Palcacocha se encuentra por debajo de lo normal informa INAIGEM tras el comunicado de la Contraloría." Nota de Prensa. Huaraz, Peru.

INEI (Instituto Nacional de Estadística e Informática). 2018. *Censos Nacionales 2017: XII de Población, VII de Vivienda y III de Comunidades Indígenas; Áncash: Resultados Definitivos.* Lima: Published by the author.

Ingold, Tim. 2000. *The Perception of the Environment: Essays on Livelihood, Dwelling, and Skill.* London: Routledge.

IPCC (Intergovernmental Panel on Climate Change). 2013. *Climate Change 2013: The Physical Science Basis; Contribution of Working Group I to the Fifth Assessment Report of the Intergovernmental Panel on Climate Change.* Vol. 1535. Edited by Thomas F. Stocker, Dahe Qin, Gian-Kasper Plattner, Melinda Tignor, Simon K. Allen, Judith Boschung, Alexander Nauels, Yu Xia, Vincent Bex, and Pauline M. Midgley. Cambridge: Cambridge University Press.

IPCC (Intergovernmental Panel on Climate Change). 2014. *Climate Change 2014: Impacts, Adaptation, and Vulnerability; Part B: Regional Aspects; Contribution of Working Group II to the Fifth Assessment Report of the Intergovernmental Panel on Climate Change.* Edited by V. R. Barros, C. B. Field, D. J. Dokken, M. D. Mastrandrea, K. J. Mach, T. E. Bilir, K. Chatterjee, K. L. Ebi, F. Estrada, R. C. Genova, B. Girma, E. S. Kissel, A. N. Levy, S. MacCracken, P. R. Mastrandrea, and L. L. White. Cambridge: Cambridge University Press.

IPCC (Intergovernmental Panel on Climate Change). 2022. *Climate Change 2022: Impacts, Adaptation, and Vulnerability; Contribution of Working Group II to the Sixth Assessment Report of the Intergovernmental Panel on Climate Change.* Edited by H.-O. Pörtner, D. C. Roberts, M. Tignor, E. S. Poloczanska, K. Mintenbeck, A. Alegría, M. Craig, S. Langsdorf, S. Löschke, V. Möller, A. Okem, and B. Rama. Cambridge: Cambridge University Press.

IPE (Instituto Peruano de Economía). 2019. "Áncash: Pobreza 2019." https://ipe.org.pe/ancash-pobreza-2019/.

Iyengar, Shalini. 2023. "Human Rights and Climate Wrongs: Mapping the Landscape of Rights-Based Climate Litigation." *Review of European, Comparative, and International Environmental Law* 32 (2): 299–309. https://doi.org/10.1111/reel.12498.

Jarvis, Brooke. 2019. "Climate Change Could Destroy His Home in Peru, So He Sued an Energy Company in Germany." *New York Times Magazine*, April 9.

Jasanoff, Sheila. 2004. "Ordering Knowledge, Ordering Society." In *States of Knowledge: The Co-Production of Science and Social Order*, edited by Sheila Jasanoff, 13–45. London: Routledge.

Jasanoff, Sheila. 2005. "Law's Knowledge: Science for Justice in Legal Settings." *American Journal of Public Health* 95 (S1): S49–58.

Jasanoff, Sheila. 2006. "Just Evidence: The Limits of Science in the Legal Process." *Journal of Law, Medicine, and Ethics* 34 (2): 328–41.

Jasanoff, Sheila. 2007. "Making Order: Law and Science in Action." In *The Handbook of Science and Technology Studies*, edited by Edward J. Hackett, Olga Amsterdamska, Michael Lynch, and Judy Wajcman, 761–86. Cambridge, MA: MIT Press.

Jasanoff, Sheila. 2010. "A New Climate for Society." *Theory, Culture, and Society* 27 (2–3): 233–53.

Jensen, Casper Bruun, and Atsuro Morita. 2017. "Introduction: Infrastructures as Ontological Experiments." *Ethnos* 82 (4): 615–26. https://doi.org/10.1080/00141844.2015.1107607.

Johnson, Lyman. 2012. "Law and Legal Theory in the History of Corporate Responsibility: Corporate Personhood." *Seattle University Law Review* 35 (4): 1135–64. https://heinonline.org/HOL/P?h=hein.journals/sealr35&i=1161.

Jurt, Christine, Maria Dulce Burga, Luis Vicuña, Christian Huggel, and Ben Orlove. 2015. "Local Perceptions in Climate Change Debates: Insights from Case Studies in the Alps and the Andes." *Climatic Change* 133 (3): 511–23.

Kaplan, Sarah. 2022. "A Melting Glacier, an Imperiled City and One Farmer's Fight for Climate Justice." *Washington Post*, August 28. https://www.washingtonpost.com/climate-environment/interactive/2022/peru-climate-lawsuit-melting-glacier/.

Kelleher, Orla. 2022. "Systemic Climate Change Litigation, Standing Rules, and the Aarhus Convention: A Purposive Approach." *Journal of Environmental Law* 34 (1): 107–34.

Kim, Claudia Junghyun, and Celeste L. Arrington. 2023. "Knowledge Production Through Legal Mobilization: Environmental Activism Against the U.S. Military Bases in East Asia." *Law and Society Review* 57 (2): 162–88. https://doi.org/10.1111/lasr.12650.

Kirsch, Stuart. 2002. "Anthropology and Advocacy: A Case Study of the Campaign Against the Ok Tedi Mine." *Critique of Anthropology* 22 (2): 175–200.

Kirsch, Stuart. 2014. "Imagining Corporate Personhood." *PoLAR: Political and Legal Anthropology Review* 37 (2): 207–17. https://doi.org/10.1111/plar.12070.

Knox, Hannah. 2015. "Thinking like a Climate." *Distinktion: Scandinavian Journal of Social Theory* 16 (1): 91–109. http://dx.doi.org/10.1080/1600910X.2015.1022565.

Knox, Hannah. 2020. *Thinking like a Climate: Governing a City in Times of Environmental Change*. Durham, NC: Duke University Press.

Kos, Andrew, Florian Amann, Tazio Strozzi, Julian Osten, Florian Wellmann, Mohammadreza Jalali, and Anja Dufresne. 2021. "The Surface Velocity Response of a Tropical Glacier to Intra and Inter Annual Forcing, Cordillera Blanca, Peru." *Remote Sensing* 13 (14): 2694.

Kotcher, John E., Teresa A. Myers, Emily K. Vraga, Neil Stenhouse, and Edward W. Maibach. 2017. "Does Engagement in Advocacy Hurt the Credibility of Scientists? Results from a Randomized National Survey Experiment." *Environmental Communication* 11 (3): 415–29. https://doi.org/10.1080/17524032.2016.1275736.

Kron, Wolfgang. 2005. "Flood Risk = Hazard • Values • Vulnerability." *Water International* 30 (1): 58–68. https://doi.org/10.1080/02508060508691837.

Kumar, Vedantha, and Will Frank. 2018. "Holding Private Emitters to Account for the Effects of Climate Change: Could a Case like Lliuya Succeed Under English Nuisance Laws?" *Carbon and Climate Law Review*, 12 (2): 110–23. https://heinonline.org/HOL/P?h=hein.journals/cclr2018&i=126.

Kysar, Douglas A. 2011. "What Climate Change Can Do About Tort Law." *Environmental Law* 41 (1): 1–71.

Lamb, William F., Giulio Mattioli, Sebastian Levi, J. Timmons Roberts, Stuart Capstick, Felix Creutzig, Jan C. Minx, Finn Müller-Hansen, Trevor Culhane, and Julia K. Steinberger. 2020. "Discourses of Climate Delay." *Global Sustainability* 3: e17. https://www.cambridge.org/core/product/7B11B722E3E3454BB6212378E32985A7.

Latour, Bruno. 1987. *Science in Action: How to Follow Scientists and Engineers Through Society*. Cambridge, MA: Harvard University Press.

Latour, Bruno. 1993. *We Have Never Been Modern*. Cambridge, MA: Harvard University Press.

Latour, Bruno. 1999. *Pandora's Hope: Essays on the Reality of Science Studies*. Cambridge, MA: Harvard University Press.

Latour, Bruno. 2004. "Why Has Critique Run Out of Steam? From Matters of Fact to Matters of Concern." *Critical inquiry* 30 (2): 225–48.

Latour, Bruno. 2010. *The Making of Law: An Ethnography of the Conseil d'Etat*. Translated by Marina Bilman and Alain Pottage. Cambridge: Polity Press.

Latour, Bruno. 2018. *Down to Earth: Politics in the New Climatic Regime*. Cambridge: Polity Press.

Latour, Bruno. 2021. *After Lockdown: A Metamorphosis*. Cambridge: Polity Press.

Leijten, Ingrid. 2019. "Human Rights v. Insufficient Climate Action: The Urgenda Case." *Netherlands Quarterly of Human Rights* 37 (2): 112–18. https://doi.org/10.1177/0924051919844375.

Li, Fabiana. 2013. "Relating Divergent Worlds: Mining, Aquifers, and Sacred Mountains in Peru." *Anthropologica* (55): 399–411.

Li, Fabiana. 2015. *Unearthing Conflict: Corporate Mining, Activism, and Expertise in Peru*. Durham, NC: Duke University Press.

Lin, Jolene, and Douglas A. Kysar, eds. 2020. *Climate Change Litigation in the Asia Pacific*. Cambridge: Cambridge University Press.

Lüning, Sebastian, Mariusz Gałka, Florencia Paula Bamonte, Felipe García-Rodríguez, and Fritz Vahrenholt. 2022. "Attribution of Modern Andean Glacier Mass Loss Requires Successful Hindcast of Pre-Industrial Glacier Changes." *Journal of South American Earth Sciences* 119: 104024. https://www.sciencedirect.com/science/article/pii/S0895981122003108.

MacKinnon, Danny. 2011. "Reconstructing Scale: Towards a New Scalar Politics." *Progress in Human Geography* 35 (1): 21–36. http://phg.sagepub.com/content/35/1/21.
Mark, Bryan G., Adam French, Michel Baraer, Mark Carey, Jeffrey Bury, Kenneth R. Young, Molly H. Polk, Oliver Wigmore, Pablo Lagos, Ryan Crumley, Jeffrey M. McKenzie, and Laura Lautz. 2017. "Glacier Loss and Hydro-Social Risks in the Peruvian Andes." *Global and Planetary Change* 159: 61–76. http://www.sciencedirect.com/science/article/pii/S0921818117301935.
Martin, Keir. 2019. "Subaltern Perspectives in Post-Human Theory." *Anthropological Theory* 20 (3): 357–82. https://doi.org/10.1177/1463499618794085.
Mattingly, Cheryl, and Jason Throop. 2018. "The Anthropology of Ethics and Morality." *Annual Review of Anthropology* 47 (1): 475–92.
Mauss, Marcel. 1985. "A Category of the Human Mind: The Notion of Person; the Notion of Self." In *The Category of the Person: Anthropology, Philosophy, History*, edited by Michael Carrithers, Steven Collins, and Steven Lukes, 1–25. Cambridge: Cambridge University Press.
Mergili, M., S. P. Pudasaini, A. Emmer, J. T. Fischer, A. Cochachin, and H. Frey. 2020. "Reconstruction of the 1941 GLOF Process Chain at Lake Palcacocha (Cordillera Blanca, Peru)." *Hydrology and Earth System Sciences* 24 (1): 93–114. https://www.hydrol-earth-syst-sci.net/24/93/2020/.
Merry, Sally Engle. 2005. "Anthropology and Activism." *PoLAR: Political and Legal Anthropology Review* 28 (2): 240–57. http://dx.doi.org/10.1525/pol.2005.28.2.240.
Miller, Clark A. 2004. "Climate Science and the Making of a Global Political Order." In *States of Knowledge: The Co-Production of Science and Social Order*, edited by Sheila Jasanoff, 46–66. London: Routledge.
Milman, Anita, John M. Marston, Sarah E. Godsey, Jessica Bolson, Holly P. Jones, and C. Susan Weiler. 2017. "Scholarly Motivations to Conduct Interdisciplinary Climate Change Research." *Journal of Environmental Studies and Sciences* 7 (2): 239–50. https://doi.org/10.1007/s13412–015–0307-z.
Minnerop, Petra. 2022. "The 'Advance Interference-Like Effect' of Climate Targets: Fundamental Rights, Intergenerational Equity, and the German Federal Constitutional Court." *Journal of Environmental Law* 34 (1): 135–62.
Mitkidis, Katerina, and Theodora N. Valkanou. 2020. "Climate Change Litigation: Trends, Policy Implications, and the Way Forward." *Transnational Environmental Law* 9 (1): 11–16.
Mouffe, Chantal. 2005. *On the Political*. London: Routledge.
Mugdan, Benno. 1899. *Die gesammelten Materialien zum Bürgerlichen Gesetzbuch für das Deutsche Reich, III. Band: Sachenrecht*. Berlin: R. v. Decker's Verlag.
Nuccitelli, Dana. 2019. "Humans and Volcanoes Caused Nearly All of Global Heating in Past 140 Years." *Guardian*, May 30. https://www.theguardian.com/environment/2019/may/30/humans-and-volcanoes-caused-nearly-all-of-global-heating-in-past-140-years.
Nugent, Ciara. 2018. "Climate Change Could Destroy This Peruvian Farmer's Home; Now He's Suing a European Energy Company for Damages." *Time*, October 5.

O'Reilly, Jessica, Cindy Isenhour, Pamela McElwee, and Ben Orlove. 2020. "Climate Change: Expanding Anthropological Possibilities." *Annual Review of Anthropology* 49: 13–29. https://doi.org/10.1146/annurev-anthro-010220-043113.

Oreskes, Naomi, and Erik M. Conway. 2010. *Merchants of Doubt: How a Handful of Scientists Obscured the Truth on Issues from Tobacco Smoke to Global Warming*. New York: Bloomsbury Press.

Orlove, Ben. 2002. *Lines in the Water: Nature and Culture at Lake Titicaca*. Berkeley: University of California Press.

Osofsky, Hari M. 2005. "The Geography of Climate Change Litigation: Implications for Transnational Regulatory Governance." *Washington University Law Quarterly* 83: 1789–1855.

Osofsky, Hari M. 2007a. "The Intersection of Scale, Science, and Law in *Massachusetts v. EPA*." *Oregon Review of International Law* 9 (2): 233–60.

Osofsky, Hari M. 2007b. "Local Approaches to Transnational Corporate Responsibility: Mapping the Role of Sub-National Climate Change Litigation." *Pacific McGeorge Global Business and Development Law Journal* 20 (1): 143–60.

Osorio Bautista, Serafín. 2013. "Acción colectiva y conflicto de intereses: El caso de la comunidad campesina de Catac." *Anthropologica* 31 (31): 43–79.

Paerregaard, Karsten. 2013. "Bare Rocks and Fallen Angels: Environmental Change, Climate Perceptions, and Ritual Practice in the Peruvian Andes." *Religions* 4 (2): 290–305.

Paerregaard, Karsten. 2020. "Communicating the Inevitable: Climate Awareness, Climate Discord, and Climate Research in Peru's Highland Communities." *Environmental Communication* 14 (1): 112–25. https://doi.org/10.1080/17524032.2019.1626754.

Paiement, Phillip. 2020. "Urgent Agenda: How Climate Litigation Builds Transnational Narratives." *Transnational Legal Theory* 11 (1–2): 121–43. https://doi.org/10.1080/20414005.2020.1772617.

Parker, Larissa, Juliette Mestre, Sébastien Jodoin, and Margaretha Wewerinke-Singh. 2022. "When the Kids Put Climate Change on Trial: Youth-Focused Rights-Based Climate Litigation Around the World." *Journal of Human Rights and the Environment* 13 (1): 64–89.

Parry, Lloyd. 2023. "Climate Sceptics Sneak Unsound Research into Peer-Reviewed Journals, Scientists Warn." *Yahoo! News*, April 12. https://uk.news.yahoo.com/climate-sceptics-sneak-unsound-research-135707994.html?guccounter=1.

Patton, Lindene, and Felicia H. Barnes. 2017. "Science and the Law: How Will Developments in Attribution Science Affect How the Law Addresses Compensation for Climate Change Effects?" In *Risk, Resilience, Inequality, and Environmental Law*, edited by Bridget M. Hutter, 147–66. Cheltenham, UK: Edward Elgar.

Peel, Jacqueline, and Hari M. Osofsky. 2015. *Climate Change Litigation: Regulatory Pathways to Cleaner Energy*. Cambridge Studies in International and Comparative Law. Cambridge: Cambridge University Press.

Peel, Jacqueline, and Hari M. Osofsky. 2017. "A Rights Turn in Climate Change Litigation?" *Transnational Environmental Law* 7 (1): 37–67.

Peel, Jacqueline, and Hari M. Osofsky. 2020. "Climate Change Litigation." *Annual Review of Law and Social Science* 16 (1): 21–38. https://doi.org/10.1146/annurev-lawsocsci-022420-122936.

Petersen, Brian, Diana Stuart, and Ryan Gunderson. 2019. "Reconceptualizing Climate Change Denial: Ideological Denialism Misdiagnoses Climate Change and Limits Effective Action." *Human Ecology Review* 25 (2): 117–42. https://www.jstor.org/stable/26964357.

Pinker, Annabel, and Penny Harvey. 2015. "Negotiating Uncertainty: Neo-Liberal Statecraft in Contemporary Peru." *Social Analysis* 59 (4): 15–31.

Poole, Deborah. 2004. "Between Threat and Guarantee: Justice and Community in the Margins of the Peruvian State." In *Anthropology in the Margins of the State*, edited by Veena Das and Deborah Poole, 35–65. Santa Fe: School of American Research Press.

Poovey, Mary. 1998. *A History of the Modern Fact: Problems of Knowledge in the Sciences of Wealth and Society*. Chicago: University of Chicago Press.

Porter, Theodore M. 1995. *Trust in Numbers: The Pursuit of Objectivity in Science and Public Life*. Princeton, NJ: Princeton University Press.

Portocarrero Rodríguez, César A. 2014. *The Glacial Lake Handbook: Reducing Risk from Dangerous Glacial Lakes in the Cordillera Blanca, Peru*. Washington, DC: United States Agency for International Development.

Preston, Brian J. 2021a. "The Influence of the Paris Agreement on Climate Litigation: Causation, Corporate Governance, and Catalyst (Part II)." *Journal of Environmental Law* 33 (2): 227–56.

Preston, Brian J. 2021b. "The Influence of the Paris Agreement on Climate Litigation: Legal Obligations and Norms (Part I)." *Journal of Environmental Law* 33 (1): 1–32.

Rabin, Robert L. 2001. "The Tobacco Litigation: A Tentative Assessment Symposium; The Changing Landscape of the Practice." *DePaul Law Review* 51 (2): 331–58. https://heinonline.org/HOL/P?h=hein.journals/deplr51&i=345.

Raiser, Thomas. 1999. "Der Begriff der juristischen Person: Eine Neubesinnung" [The Concept of the Juridical Person. A Reconsideration]. *Archiv für die civilistische Praxis* 199 (1–2): 104–44. www.jstor.org/stable/40995773.

Rasmussen, Mattias Borg. 2015. *Andean Waterways: Resource Politics in Highland Peru*. Seattle: University of Washington Press.

Rodríguez-Garavito, César, ed. 2022. *Litigating the Climate Emergency: How Human Rights, Courts, and Legal Mobilization Can Bolster Climate Action*. Cambridge: Cambridge University Press.

Rogers, Nicole. 2013. "Climate Change Litigation and the Awfulness of Lawfulness." *Alternative Law Journal* 38 (1): 20–24. https://journals.sagepub.com/doi/abs/10.1177/1037969X1303800105.

Rogers, Nicole. 2020. *Law, Fiction, and Activism in a Time of Climate Change*. New York: Routledge.

Rose, Nikolas S. 1996. *Inventing Our Selves: Psychology, Power, and Personhood*. Cambridge Studies in the History of Psychology. Cambridge: Cambridge University Press.

Rowlatt, Justin. 2020. "Greta Thunberg: Climate Change 'as Urgent' as Coronavirus." BBC News. Updated June 22. https://www.bbc.co.uk/news/science-environment-53100800.

Roy, Eleanor Ainge. 2017. "New Zealand Gives Mount Taranaki Same Legal Rights as a Person." *Guardian*, December 22. https://www.theguardian.com/world/2017/dec/22/new-zealand-gives-mount-taranaki-same-legal-rights-as-a-person.

RWE. 2017. "Geschäftsbericht 2016." Accessed May 22, 2020. https://www.group.rwe/investor-relations/finanzberichte-praesentationen-videos/finanzberichte.

RWE. 2022. *Sustainability Report 2021*. Essen: Published by the author. https://www.rwe.com/-/media/RWE/documents/09-verantwortung-nachhaltigkeit/cr-berichte/EN/cr-report-2021.pdf.

RWE. 2024. "Voller Energie in eine grüne Zukunft: Geschäftsbericht 2023." March 14. https://www.rwe.com/-/media/RWE/documents/05-investor-relations/finanzkalender-und-veroeffentlichungen/2023-Q4/2024-03-14-rwe-geschaeftsbericht-2023.pdf.

Sahlins, Marshall. 2013. *What Kinship Is—And Is Not*. Chicago: University of Chicago Press.

Samuel, Sigal. 2019. "This Country Gave All Its Rivers Their Own Legal Rights." *Vox*, August 18.

Sapignoli, Maria. 2017. "'Bushmen' in the Law: Evidence and Identity in Botswana's High Court." *PoLAR: Political and Legal Anthropology Review* 40 (2): 210–25. https://doi.org/10.1111/plar.12216.

Sato, Misato, Glen Gostlow, Catherine Higham, Joana Setzer, and Frank Venmans. 2024. "Impacts of Climate Litigation on Firm Value." *Nature Sustainability* 7: 1461–68. https://doi.org/10.1038/s41893-024-01455-y.

Sawyer, Suzana. 2022. *The Small Matter of Suing Chevron*. Durham, NC: Duke University Press.

Sayre, Nathan F. 2012. "The Politics of the Anthropogenic." *Annual Review of Anthropology* 41 (1): 57–70. http://www.annualreviews.org/doi/abs/10.1146/annurev-anthro-092611-145846.

Schauer, Frederick. 2022. *The Proof: Uses of Evidence in Law, Politics, and Everything Else*. Cambridge, MA: Belknap Press of Harvard University Press.

Schauwecker, S., M. Rohrer, D. Acuña, A. Cochachin, L. Dávila, H. Frey, C. Giráldez, J. Gómez, C. Huggel, and M. Jacques-Coper. 2014. "Climate Trends and Glacier Retreat in the Cordillera Blanca, Peru, Revisited." *Global and Planetary Change* 119: 85–97.

Scheper-Hughes, Nancy. 1995. "The Primacy of the Ethical: Propositions for a Militant Anthropology." *Current Anthropology* 46 (3): 409–40.

Schneider, D., C. Huggel, A. Cochachin, S. Guillén, and J. García. 2014. "Mapping Hazards from Glacier Lake Outburst Floods Based on Modelling of Process Cascades at Lake 513, Carhuaz, Peru." *Advances in Geosciences* 35: 145–55. http://www.adv-geosci.net/35/145/2014/.

Setzer, Joana, and Lisa Benjamin. 2020. "Climate Change Litigation in the Global South: Filling in Gaps." AJIL *Unbound* 114: 56–60.

Setzer, Joana, and Catherine Higham. 2025. *Global Trends in Climate Change Litigation: 2025 Snapshot*. London: Grantham Research Institute on Climate Change and the Environment and Centre for Climate Change Economics and Policy, London School of Economics and Political Science.

Setzer, Joana, and Lisa C. Vanhala. 2019. "Climate Change Litigation: A Review of Research on Courts and Litigants in Climate Governance." *WIREs Climate Change* 10 (3): e580. https://onlinelibrary.wiley.com/doi/abs/10.1002/wcc.580.

Somos-Valenzuela, M. A., R. E. Chisolm, D. S. Rivas, C. Portocarrero, and D. C. McKinney. 2016. "Modeling a Glacial Lake Outburst Flood Process Chain: The Case of Lake Palcacocha and Huaraz, Peru." *Hydrology and Earth System Science* 20 (6): 2519–43. http://www.hydrol-earth-syst-sci.net/20/2519/2016/.

SourceMaterial. 2022. "'Battle of Science' Rages over Peru Glacier." June 3. https://www.source-material.org/battle-of-science-rages-over-peru-glacier/.

Spencer, Roy W., and William D. Braswell. 2011. "On the Misdiagnosis of Surface Temperature Feedbacks from Variations in Earth's Radiant Energy Balance." *Remote Sensing* 3 (8): 1603–13.

Star, Susan Leigh, and Martha Lampland. 2009. "Reckoning with Standards." In *Standards and their Stories: How Quantifying, Classifying and Formalizing Practices Shape Everyday Life*, edited by Martha Lampland and Susan Leigh Star, 3–24. Ithaca, NY: Cornell University Press.

Starr, June, and Jane F. Collier. 1989. "Introduction: Dialogues in Legal Anthropology." In *History and Power in the Study of Law: New Directions in Legal Anthropology*, edited by June Starr and Jane F. Collier, 1–28. Ithaca, NY: Cornell University Press.

Stehr, Nico, and Bernd Weiler. 2008. *Who Owns Knowledge? Knowledge and the Law*. New Brunswick, NJ: Transaction.

Stengers, Isabelle. 2005. "The Cosmopolitical Proposal." In *Making Things Public: Atmospheres of Democracy*, edited by Bruno Latour and Peter Weibel, 994–1003. Cambridge, MA: MIT Press.

Stensrud, Astrid B. 2016a. "Climate Change, Water Practices, and Relational Worlds in the Andes." *Ethnos* 81 (1): 75–98. http://dx.doi.org/10.1080/00141844.2014.929597.

Stensrud, Astrid B. 2016b. "'It Seems Like a Lie': The Everyday Politics of World-Making in Contemporary Peru." In *Critical Anthropological Engagements in Human Alterity and Difference*, edited by Bjørn Enge Bertelsen and Synnøve Bendixsen, 253–72. Cham, Switzerland: Springer International.

Stensrud, Astrid B. 2019a. "Water as Resource and Being: Water Extractivism and Life Projects in Peru." In *Indigenous Life Projects and Extractivism: Ethnographies from South America*, edited by Cecilie Vindal Ødegaard and Juan Javier Rivera Andía. Cham, Switzerland: Palgrave Macmillan.

Stensrud, Astrid B. 2019b. "'You Cannot Contradict the Engineer': Disencounters of Modern Technology, Climate Change, and Power in the Peruvian Andes." *Critique of Anthropology* 39 (4). https://journals.sagepub.com/doi/abs/10.1177/0308275X18821164.

Stone, Christopher D. 1972. "Should Trees Have Standing—Toward Legal Rights for Natural Objects." *Southern California Law Review* 45 (2): 450–501.

Strathern, Marilyn. 2018. "Opening Up Relations." In *A World of Many Worlds*, edited by Marisol de la Cadena and Mario Blaser, 23–52. Durham, NC: Duke University Press.

Stuart-Smith, Rupert, Friederike Otto, Aisha Saad, Gaia Lisi, Petra Minnerop, Kristian Cedervall Lauta, Kristin van Zwieten, and Thom Wetzer. 2021a. "Filling the Evidentiary Gap in Climate Litigation." *Nature Climate Change* 11 (8): 651–55. https://doi.org/10.1038/s41558-021-01086-7.

Stuart-Smith, Rupert, Gerard Roe, Sihan Li, and Myles Allen. 2021b. "Increased Outburst Flood Hazard from Lake Palcacocha Due to Human-Induced Glacier Retreat." *Nature Geoscience* 14 (2): 85–90. https://doi.org/10.1038/s41561-021-00686-4.

Stuart-Smith, Rupert, Friederike Otto, and Thom Wetzer. 2022. "Liability for Climate Change Impacts: The Role of Climate Attribution Science." In *Corporate Responsibility and Liability in Relation to Climate Change*, edited by Elbert R. De Jong, 1–27. Cambridge: Intersentia.

Stuart-Smith, Rupert, Gerard Roe, Sihan Li, and Myles Allen. 2023. "Comment on 'Attribution of Modern Andean Glacier Mass Loss Requires Successful Hindcast of Pre-Industrial Glacier Changes,' by Sebastian Lüning et al." *Journal of South American Earth Sciences*. http://dx.doi.org/10.2139/ssrn.4410943.

Supran, Geoffrey, and Naomi Oreskes. 2021. "Rhetoric and Frame Analysis of ExxonMobil's Climate Change Communications." *One Earth* 4 (5): 696–719. https://www.sciencedirect.com/science/article/pii/S2590332221002335.

Tănăsescu, Mihnea. 2020. "Rights of Nature, Legal Personality, and Indigenous Philosophies." *Transnational Environmental Law* 9 (3): 429–53. https://www.cambridge.org/core/product/398F646381C9733DE5789024BF5F9962.

Tănăsescu, Mihnea. 2022. *Understanding the Rights of Nature: A Critical Introduction*. Bielefeld, Germany: Transcript Verlag.

Thiranagama, Sharika. 2019. "Respect Your Neighbor as Yourself: Neighborliness, Caste, and Community in South India." *Comparative Studies in Society and History* 61 (2): 269–300.

Tollefson, Jeff. 2021. "Top Climate Scientists Are Sceptical That Nations Will Rein in Global Warming." *Nature* 599 (7883): 22–24.

Toussaint, Patrick. 2021. "Loss and Damage and Climate Litigation: The Case for Greater Interlinkage." *Review of European, Comparative, and International Environmental Law* 30 (1): 16–33.

Tsing, Anna Lowenhaupt. 2005. *Friction: An Ethnography of Global Connection*. Princeton, NJ: Princeton University Press.

Ule, Carl Hermann, Hans-Werner Laubinger, and Ulrich Repkewitz. 2014. *Bundes-Immissionsschutzgesetz*. Köln: Carl Heymanns Verlag.

University of Oxford. 2021. "Severe flood threat caused by climate change—landmark study." February 4. https://www.ox.ac.uk/news/2021-02-04-severe-flood-threat-caused-climate-change-landmark-study.

Urbina, Laura. 2017. "Áncash: Contraloría pide reforzar laguna Palcacocha." *El Comercio*, May 2. https://elcomercio.pe/peru/ancash/ancash-contraloria-pide-reforzar-laguna-palcacocha-418157-noticia/.

Valverde, Mariana. 2005. "Authorizing the Production of Urban Moral Order: Appellate Courts and Their Knowledge Games." *Law and Society Review* 39 (2): 419–56. https://doi.org/10.1111/j.0023-9216.2005.00087.x.

Vanhala, Lisa. 2020. "Coproducing the Endangered Polar Bear: Science, Climate Change, and Legal Mobilization." *Law and Policy* 42 (2): 105–24. https://doi.org/10.1111/lapo.12144.

Vanhala, Lisa. 2022a. "Environmental Legal Mobilization." *Annual Review of Law and Social Science* 18 (1): 101–17. https://doi.org/10.1146/annurev-lawsocsci-050520-104423.

Vanhala, Lisa. 2022b. "The Social and Political Life of Climate Change Litigation." In Rodríguez-Garavito, *Litigating the Climate Emergency*, 84–94.

Van Vleet, Krista. 2003. "Partial Theories: On Gossip, Envy, and Ethnography in the Andes." *Ethnography* 4 (4): 491–519. http://www.jstor.org/stable/24047931.

Verheyen, Roda. 2005. *Climate Change Damage and International Law: Prevention Duties and State Responsibility*. Developments in International Law 54. Leiden: Martinus Nijhoff.

Verheyen, Roda. 2015. "Loss and Damage Due to Climate Change: Attribution and Causation-Where Climate Science and Law Meet." *International Journal of Global Warming* 8 (2): 158–169.

Verheyen, Roda, and Johannes Franke. 2023. "Climate Change Litigation: A Reference Area for Liability." In *Corporate Liability for Transboundary Environmental Harm: An International and Transnational Perspective*, edited by Peter Gailhofer, David Krebs, Alexander Proelss, Kirsten Schmalenbach, and Roda Verheyen, 353–418. Cham, Switzerland: Springer International.

Vilímek, Vít, Jan Klimeš, Adam Emmer, and Jan Novotný. 2014. "Natural Hazards in the Cordillera Blanca of Peru During the Time of Global Climate Change." In *Landslide Science for a Safer Geoenvironment*, 261–66. Cham, Switzerland: Springer.

Vilímek, Vít, Marco Luyo Zapata, Jan Klimeš, Zdeněk Patzelt, and Nelson Santillán. 2005. "Influence of Glacial Retreat on Natural Hazards of the Palcacocha Lake Area, Peru." *Landslides* 2 (2): 107–15. https://doi.org/10.1007/s10346-005-0052-6.

Viveiros de Castro, Eduardo. 2012. "Immanence and Fear: Stranger-Events and Subjects in Amazonia." *HAU: Journal of Ethnographic Theory* 2 (1): 27–43.

Von Schnitzler, Antina. 2014. "Performing Dignity: Human Rights, Citizenship, and the Techno-Politics of Law in South Africa." *American Ethnologist* 41 (2): 336–50.

Von Storch, Lilian, Lukas Ley, and Jing Sun. 2021. "New Climate Change Activism: Before and After the Covid-19 Pandemic." *Social Anthropology* 29 (1): 205.

Von Ungern-Sternberg, Antje. 2013. "Normative Wirkungen von Präjudizien nach der Rechtsprechung des Bundesverfassungsgerichts" (Normative effects of precedents according to the jurisprudence of the Federal Constitutional Court). *Archiv des Öffentlichen Rechts* 138 (1): 1–59. http://www.jstor.org/stable/44318226.

Walker, Thomas W. 2008. "Who Is My Neighbor? An Invitation to See the World with Different Eyes." In *Global Neighbors: Christian Faith and Moral Obligation in Today's Economy*, edited by Douglas A. Hicks and Mark Valeri, 1–15. Grand Rapids, MI: Eerdmans.

Walker-Crawford, Noah. 2021. "The Moral Climate of Melting Glaciers: Andean Claims for Justice at the Paris Climate Change Summit." In *The Anthropocene of Weather and Climate: Ethnographic Contributions to the Climate Change Debate*, edited by Paul Sillitoe. New York: Berghahn.

Walker-Crawford, Noah. 2023. "Climate Change in the Courtroom: An Anthropology of Neighborly Relations." *Anthropological Theory* 21 (1): 76–99. https://doi.org/10.1177/14634996221138338.

Walker-Crawford, Noah. 2024. "Justice for Glaciers." *Journal of Legal Anthropology* 8 (1): 1–19.

Watts, Jonathan. 2017. "Germany's Dirty Coalmines Become the Focus for a New Wave of Direct Action." *Guardian*, November 8. https://www.theguardian.com/environment/2017/nov/08/germanys-dirty-coalmines-become-the-focus-for-a-new-wave-of-direct-action.

Wegner, Steven A. 2014. *Lo Que el Agua se Llevó: Consecuencias y Lecciones del Aluvión de Huaraz de 1941*. Lima: Ministerio del Ambiente.

Welker, Marina. 2014. *Enacting the Corporation: An American Mining Firm in Post-Authoritarian Indonesia*. Berkeley: University of California Press.

Wewerinke-Singh, Margaretha. 2023. "The Rising Tide of Rights: Addressing Climate Loss and Damage Through Rights-Based Litigation." *Transnational Environmental Law* 12 (3): 537–66. https://www.cambridge.org/core/product/5FB023532C56BB675D110B8ECE5B3910.

Wewerinke-Singh, Margaretha, and Ashleigh McCoach. 2021. "The State of the Netherlands v. Urgenda Foundation: Distilling Best Practice and Lessons Learnt for Future Rights-Based Climate Litigation." *Review of European, Comparative, and International Environmental Law* 30 (2): 275–83.

Wewerinke-Singh, Margaretha, and Zoe Nay. 2023. "Climate Change as a Children's Rights Crisis: Procedural Obstacles in International Rights-Based Climate Litigation." In *European Yearbook on Human Rights 2022*, edited by Philip Czech, Lisa Heschl, Karin Lukas, Manfred Nowak, and Gerd Oberleitner, 647–76. Cambridge: Intersentia.

Whitington, Jerome. 2016. "What Does Climate Change Demand of Anthropology?" *PoLAR: Political and Legal Anthropology Review* 39 (1): 7–15.

Winter, Gerd. 2022. "The Intergenerational Effect of Fundamental Rights: A Contribution of the German Federal Constitutional Court to Climate Protection." *Journal of Environmental Law* 34 (1): 209–21.

Wonneberger, Anke. 2023. "Climate Change Litigation in the News: Litigation as Public Campaigning Tool to Legitimize Climate-Related Responsibilities and Solutions." *Social Movement Studies* 23 (1): 94–112.

Wonneberger, Anke, and Rens Vliegenthart. 2021. "Agenda-Setting Effects of Climate Change Litigation: Interrelations Across Issue Levels, Media, and Politics in the Case of Urgenda Against the Dutch Government." *Environmental Communication* 15 (5): 699–714.

Wynne, Brian. 2010. "Strange Weather, Again." *Theory, Culture, and Society* 27 (2–3): 289–305. https://doi.org/10.1177/0263276410361499.

Yauri Montero, Marcos. 2000. *Leyendas Ancashinas*. Lima: Lerma Gómez eirl.

Young, Robert, Michael Faure, and Paul Fenn. 2004. "Causality and Causation in Tort Law." *International Review of Law and Economics* 24 (4): 507–23. http://www.sciencedirect.com/science/article/pii/S0144818805000086.

Zabiliūtė, Emilija. 2020. "Ethics of Neighborly Intimacy Among Community Health Activists in Delhi." *Medical Anthropology* 40 (1): 20–34. https://doi.org/10.1080/01459740.2020.1764550.

Index

Page numbers in italics indicate figures.

www.ingramcontent.com/pod-product-compliance
Lightning Source LLC
Chambersburg PA
CBHW032348280326
41935CB00008B/499
9781478033172